Emergency Services Sector Protection and Homeland Security

ALSO IN THE HOMELAND SECURITY SERIES

Chemical Infrastructure Protection and Homeland Security
Commercial Facilities Protection and Homeland Security
Communication Sector Protection and Homeland Security
Dam Sector Protection and Homeland Security
Defense Industrial Base Protection and Homeland Security
Energy Infrastructure Protection and Homeland Security
Financial Services Sector Protection and Homeland Security
Food Supply Protection and Homeland Security
Government Facilities Protection and Homeland Security
Information Technology Protection and Homeland Security
Nuclear Infrastructure Protection and Homeland Security
Transportation Protection and Homeland Security
Water Infrastructure Protection and Homeland Security

Emergency Services Sector Protection and Homeland Security

Frank R. Spellman

Lanham • Boulder • New York • London

An imprint of The Rowman & Littlefield Publishing Group, Inc.
4501 Forbes Boulevard, Suite 200, Lanham, Maryland 20706
www.rowman.com

86-90 Paul Street, London EC2A 4NE

British Library Cataloguing in Publication Information Available

Library of Congress Cataloging-in-Publication Data

Names: Spellman, Frank R., author.
Title: Emergency services sector protection and homeland security / Frank R. Spellman.
Description: Lanham : Bernan Press, [2023] | Series: Homeland security series | Includes bibliographical references and index.
Identifiers: LCCN 2023023286 (print) | LCCN 2023023287 (ebook) | ISBN 9781641433969 (paperback ; alk. paper) | ISBN 9781641433976 (epub)
Subjects: LCSH: Emergency management—United States. | National security—United States. | Internal security—United States.
Classification: LCC HV551.3 .S65 2023 (print) | LCC HV551.3 (ebook) | DDC 363.34/80973—dc23/eng/20230626
LC record available at https://lccn.loc.gov/2023023286
LC ebook record available at https://lccn.loc.gov/2023023287

∞™ The paper used in this publication meets the minimum requirements of American National Standard for Information Sciences—Permanence of Paper for Printed Library Materials, ANSI/NISO Z39.48-1992.

Contents

Preface

As the fifteenth volume of this well-received and highly acclaimed series on critical infrastructure and homeland security, *Emergency Services Sector Protection and Homeland Security* is an eye-opening account and an important reference source of a diverse and complex sector. This book was designed and written to serve and advise U.S. financial planners, architects, project designers, engineers, communications technicians, law enforcement and security specialists, emergency service and response personnel, managers, and superintendents and/or supervisors and responsible-managers-in-charge of protecting the multifaceted nature of critical infrastructure in the United States. When I mentioned to a colleague of mine that I was in the process of writing another volume of this series, and I described the topic to be covered, she was surprised that I would embark on such a mind-numbing and difficult task.

The emergency services sector (ESS) is crucial to all critical infrastructure sectors, as well as to the American public. As its operations provide the first line of defense for nearly all critical infrastructure sectors, a failure or disruption of the ESS could result in significant harm or loss of life, major public health issues, longer-term economic loss, and cascading disruptions of other critical infrastructure. Facilities within the sector operate on the principle of limited open public access (a key factor for terrorists), meaning that the general public can move freely without the deterrent of highly visible or obstructive security barriers.

The bottom line: in some cases and locations, the ESS is a wide-open, easy target for terrorists.

This book is organized to simplify and present, in a logical and sequential manner, a discussion not only of the entities or units (i.e., law enforcement, fire and rescue services, emergency medical services, emergency management, public works) comprising the ESS in the United States but also many of the security measures employed to protect the various entities, personnel, and equipment involved.

Let's face today's reality: those who want quick answers to complicated questions—to help employers and employees handle security threats—must be prepared to meet and deal with the threat of terrorism on a 24/7 basis. It is important to point out that this book does not discuss or focus on security concerns related to natural disasters. On the contrary, the focus here is on the security aspects in the design of systems needed to deal robustly with possible sources of disruption, specifically from malevolent acts. Moreover, the focus includes the added dimension of preventing misuse and malicious behavior. In the post-9/11 world, the possibility of ESS infrastructure terrorism—the use of toxic substances, weapons, and/or cyber intrusion to cause devastating damage to the ESS infrastructure and its associated subsectors, along with, literally, its cascading effects—is very real. Thus, the need is clear and real, and so are the format and guidelines presented in this text to improve the protection and resilience of the ESS infrastructure.

This book describes the sector/subsector-wide process required to identify and prioritize assets, assess risk in the sector, implement protective programs and resilience strategies, and measure their effectiveness. This book and the complete sixteen volumes (upgraded from the original fourteen volumes) of the critical infrastructure sector series were written as a result of 9/11 to address these concerns. It is important to point out that our ESS infrastructure (as is the case with the other fifteen critical infrastructures) cannot be made absolutely immune to all possible intrusions or attacks; thus, it takes a concerted, well-thought-out effort to incorporate security upgrades in the retrofitting of existing systems and careful security planning for all new facility infrastructure components. These upgrades or design features need to address issues of monitoring, response, critical infrastructure redundancy, and recovery to minimize risk to the facility infrastructure. However, based on personal experience, none of these approaches is or can be effective unless ESS staff members at all levels of the chain of command are cognizant of the threats.

Emergency Services Sector Protection and Homeland Security presents commonsense methodologies in a straightforward, almost blunt manner. Why so blunt? When dealing with protecting our critical infrastructure, I am always blunt . . . aren't you? Think about it: at this particular time, when dealing with the security of workers, family members, citizens, and society in general—actually, with our very way of life—politically correct presentations on security might be the norm, might be expected, and might be demanded by the far left, but my view is that there is nothing normal or subtle about killing thousands of innocent people; mass murders certainly should not be expected; the right and need to communicate and the right to live in a free and safe environment are reasonable demands.

This text is accessible to those who have no experience with, or knowledge of, the emergency services sector. If you work through the text systematically, you will gain an understanding of the challenge of domestic preparedness—that is, an immediate need for a heightened state of awareness of the present threat facing the ESS and critical infrastructure sectors as potential terrorist targets. Moreover, you will gain knowledge of security principles and measures that can be implemented, not only adding a critical component to your professional knowledge but also giving you the tools needed to combat terrorism in the homeland—our homeland—both by outsiders and insiders.

One final word to readers: this book is written in the conversational, engaging, and reader-friendly style that is the author's trademark. Why? Well, when demonstrating how one—or anyone—can write about the U.S. emergency services sector, when it is so deep, tall, wide, and all-encompassing, and almost indefinable in terms of total reach, scope, extent, etc., etc., etc., I never apologize for attempting to communicate.

Acronyms and Abbreviations

AASHTO	American Association of State Highway and Transportation Officials
ANSI	American National Standards Institute
APT	advanced persistent threat
ASSE	American Society of Sanitary Engineers
ASTM	American Society for Testing and Materials
BMI	base interface module
CAD	computer-aided dispatch
CBRN	chemical, biological, radiological, or nuclear
CBRNE	chemical, biological, radiological, nuclear, or explosive
CCD	charge-coupled device
CCTV	closed-circuit television
CMS	critical manufacturing sector
CNCI	Comprehensive National Cybersecurity Initiative
CSB	Chemical Safety Board
CSI	crime scene investigation
DBT	design basis threat
DCS	distributed control system
DDoS	distributed denial of service
DEQ	department of environmental quality
DHS	Department of Homeland Security
DIB	defense industrial base
DMZ	demilitarized zone
DNDO	Domestic Nuclear Detection Office
DNS	domain name system
DOD	Department of Defense
DOE	Department of Energy
EAP	emergency action procedure

EAS	Emergency Alert System
EMS	emergency medical services
EPA	Environmental Protection Agency
EOC	emergency operations center
ERP	emergency response plan
ES	emergency services
ESMR	enhanced special mobile radio
ESP	emergency services plan
ESS	emergency services sector
FBI	Federal Bureau of Investigation
FEMA	Federal Emergency Management Agency
FlaWARN	Florida's Water/Wastewater Agency Response Network
FMEA	failure mode and effects
GAO	Government Accountability Office
GIS	geographic information system
GPS	Global Positioning System
ICS-CERT	Industrial Control Systems Cyber Emergency Response Team
IDS	intrusion detection system
IED	improvised explosive device
IEEE	Institute of Electrical and Electronic Engineers
IoT	internet of things
IP	Internet Protocol
IR	infrared
ISO	independent system operator
IT	information technology
HAZMAT	hazardous materials
HAZOP	hazard and operability
HMI	human-machine interface
HPH	health care and public health
K-9	canine
LAN	local area network
LLE	local law enforcement
LMR	land mobile radio
MTU	master terminal unit
NC	normally closed
NCIC	National Crime Information Center
NCSD	National Cybersecurity Division
NDWAC	National Drinking Water Advisor Council
NEC	National Electrical Codes
NEMSIS	National EMS Information System
NFPA	National Fire Protection Association
NIMS	National Incident Management System
NIPC	National Infrastructure Protection Center

NO	normally open
OPSEC	operations security
OSHA	Occupational Safety and Health Administration
P&ID	piping and instrumentation diagram
PDA	personal digital assistant
PFD	process flow diagram
PIR	passive infrared
PLC	programmable logic controller
POTS	plain old telephone service
PRRA	pre-removal risk assessment
PSAP	public services answering points
PSC&C	public safety communications and coordination
PSM	process safety management
PTZ	pan, tilt, zoom
RF	radio frequency
RMP	risk management planning
ROE	rules of engagement
RPTB	Response Protocol Toolbox
RTU	remote terminal unit
SaaS	software-as-a-service
SCADA	supervisory control and data acquisition
SQL	Structured Query Language
SWAT	special weapons and tactics
Tbps	terabits per second
TDoS	telephone denial of service
UAS	unmanned aircraft system
US-CERT	U.S. Computer Emergency Response Team
USFA	U.S. Fire Administration
UV	ultraviolet
VA	vulnerability assessment
VMD	video motion detection
WAP	wireless access point
WMD	weapon of mass destruction
WPS	Wireless Priority Service
WSWG	Water Security Working Group
XML	Extensible Markup Language

Chapter 1

Introduction

EMERGENCY SERVICES SECTOR

The emergency services sector (ESS) is one of the sixteen critical infrastructure sectors. The ESS is a community of millions of highly skilled and trained personnel, along with the physical and cyber resources that provide a wide range of prevention, preparedness, response, and recovery services in both day-to-day operations and incident response. The ESS represents the nation's first line of defense in the prevention and mitigation of risk from both intentional and unintentional man-made incidents, as well as from natural disasters. The ESS supports each of the other fifteen critical infrastructure sectors and assists a range of organizations and communities in maintaining public safety, security, and confidence in the government by performing lifesaving operations, protecting property and the environment, assisting communities impacted by disasters, and aiding recovery from emergencies.

SECTOR OVERVIEW

The emergency services sector includes geographically distributed facilities and equipment in both paid and volunteer capacities organized primarily at the federal, state, local, tribal, and territorial levels of government. These include city police departments and fire stations, county sheriff's offices, Department of Defense police and fire departments, and town public works departments. The ESS also includes private-sector resources such as industrial fire departments, private security organizations, and private emergency medical services providers (DHS, 2019).

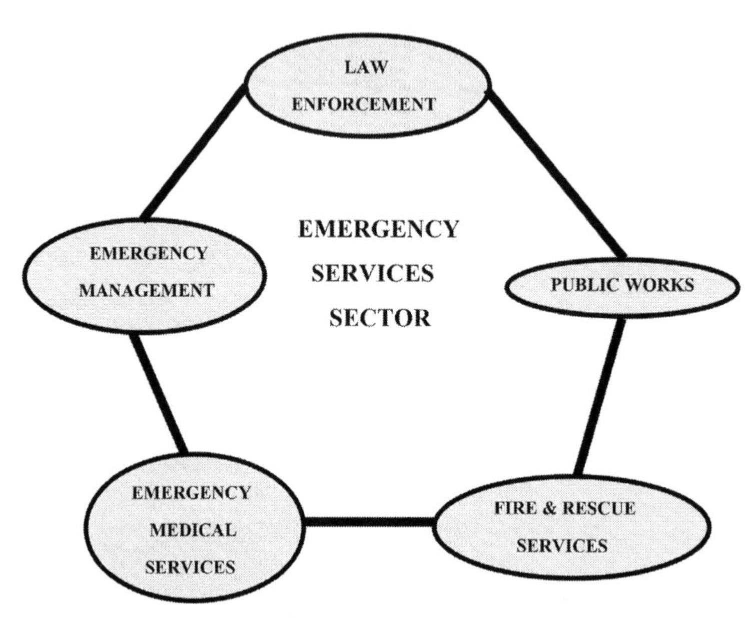

Figure 1.1. Emergency Services Sector

The mission of the ESS is straightforward: to save lives, protect property and the environment, assist communities impacted by disasters, and aid recovery during emergencies.

The five distinct disciplines that make up the ESS are shown in figure 1.1—law enforcement, fire and rescue services, emergency medical services, emergency management, and public works. These disciplines encompass a wide range of emergency response functions and specialized services provided by individual personnel and by teams.

Fire and Rescue Services

The United States Fire Administration (USFA) is an entity of the U.S. Department of Homeland Security's Federal Emergency Management Agency (FEMA). The USFA defines "Fire and Rescue Services" as an entity encompassing a variety of capabilities and fire service–related organizations. These include structural and wildland firefighting, technical rescue services, and emergency medical services, as well as state fire marshal's offices and equivalent agencies, firefighter associates, and other natural-level entities.

What is the difference between a firefighter and a fire marshal? Good question.

A *firefighter* is a rescuer extensively trained in firefighting and rescue techniques. A *fire marshal* is a fire official who has been lawfully appointed and charged with statutory responsibilities and duties for fire prevention.

Let's look at some statistics related to U.S. Fire and Rescue Services (FEMA, 2019):

- There are more than 1,000,000 career, volunteer, and paid-per-call firefighters.
- There are more than 27,000 firefighters registered with the U.S. Fire Administration.
- There are more than 151,000 staff and non-firefighting personnel.
- There are almost 58,000 fire stations.
- Approximately 69 percent of fire departments have one fire station.
- 14 percent of fire departments have three or more stations.

Fire departments provide the following specialized services:

- Basic and advanced life support
- Airport/aviation
- In-house departmental training
- EMS nontransport response
- EMS ambulance transport
- Fire inspection/code enforcement
- Fire inspection/cause determination
- Fireboats
- Hazardous materials (HAZMAT) teams
- Juvenile fire-setter intervention programs
- Public education
- Technical/specialized rescue
- Vehicle extrication
- Wildlife/wildland urban interface

Law Enforcement

Law enforcement is a term that describes the individuals and agencies responsible for enforcing laws and maintaining public order and public safety. Law enforcement includes the prevention, detection, and investigation of crimes and the apprehension and detention of individuals suspected of law violations. The law-enforcement community consists of federal and state, local, tribal, and territorial law-enforcement agencies, court systems, correctional institutions, and private-sector security agencies.

At the present time there are seventy-three federal law-enforcement agencies (e.g., the U.S. Secret Service, the Drug Enforcement Administration, U.S. Park Police, the Office of Inspectors General, and so forth). State, local, tribal, and territorial law-enforcement agencies consist of local police departments, sheriff's offices, primary state law-enforcement agencies, special jurisdiction agencies, and other agencies. The private-sector security companies and other protective-service professions consist of private-sector security guards, transit and rail police, and others. Several of the specialized law-enforcement capabilities include:

- Aviation unit
- Bomb squad/explosive unit
- Canine (K-9) unit
- Crime scene investigation (CSI)
- Crisis negotiation
- Hazardous materials (HAZMAT)
- Marine and port unit
- Patrol or strike team
- Public safety dive team
- Riot/crowd control
- Search and rescue
- Special weapons and tactics (SWAT)

Emergency Medical Services

Emergency medical services (EMS) provide pre-hospital emergency medical care via a system of coordinated response, including multiple people and agencies (see figure 1.2). EMS practitioners may provide both basic and advanced medical care at the scene of an emergency and en route to the hospital.

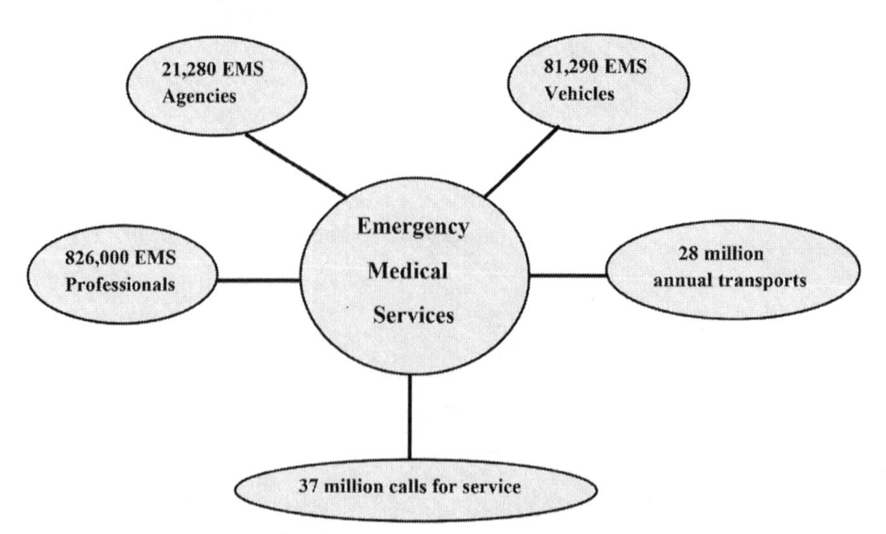

Figure 1.2. Emergency Medical Services

No matter the location, the essential components of an EMS system remain the same. The types of EMS agencies include (FEMA, 2019):

- Fire department based
- Private, not hospital based
- State and local government, not fire based
- Hospital based
- Emergency medical dispatch
- Tribal
- Other EMS agency

Public Works

There are approximately nineteen thousand municipalities of varying sizes in the United States. Each community has common needs that must be met through the provision of public works services (wastewater treatment, drinking water, utilities, trash collection, etc.; see figure 1.3). In this text, *public works* is defined as the combination of physical assets, management practices, policies, and personnel necessary for the government to provide and sustain structures and services essential to the welfare and acceptable quality of life of its citizens. From an emergency management perspective (and germane to the goals of homeland security), public works is an integral

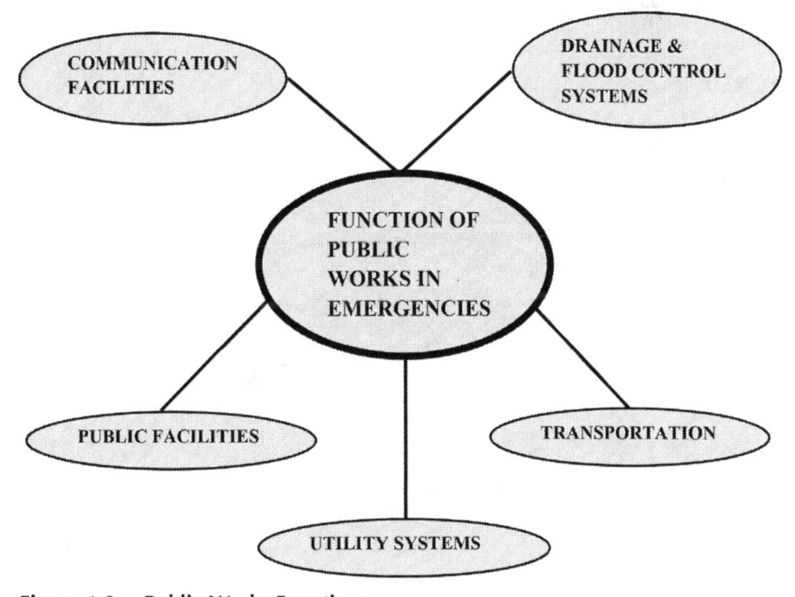

Figure 1.3. Public Works Functions

component of a jurisdiction's emergency planning and response efforts (both pre- and post-event) (FEMA, 2019).

Specialized services provided by public works include:

- Construction management
- Emergency management
- Engineering and technology
- Facilities
- Fleet services
- Grounds and urban forestry
- Solid waste management
- Water management
- Transportation management
- Utility and public right-of-way

Emergency Management

Reducing vulnerability to hazards and coping with disasters is the managerial function of emergency management. This managerial function is typically undertaken by a designated *emergency manager*, who has a working knowledge of all the basic tenets of emergency management, including mitigation, preparedness, response and recovery, and the knowledge, skills, and ability to effectively manage a locality's emergency management program. Emergency management specialists and emergency management directors (~9,840 in the United States) plan for, coordinate, and manage response efforts (FEMA, 2019).

In addition to the foundational capabilities of the disciplines, federal, state, local, territorial, tribal, and private-sector assets, networks, and systems provide specialized emergency services (see figure 1.4) through individual personnel and teams. These specialized capabilities may be found in one or more various disciplines, depending on the jurisdiction.

Sector Man-Made Threats and Capabilities

The evolving and oftentimes unpredictable nature of man-made threats is an ever-evolving challenge for response efforts. Serving and protecting the public requires ESS personnel to maintain a high level of threat awareness and associated degrees of response, as well as the capacity to respond to an increasing number of complex challenges. Evolving threats include asymmetrical attack incidents, such as active shooter and improvised explosive device (IED) incidents; biological agents and infectious diseases occurrences (deliberate releases—weaponized with malicious intent), such as Ebola,

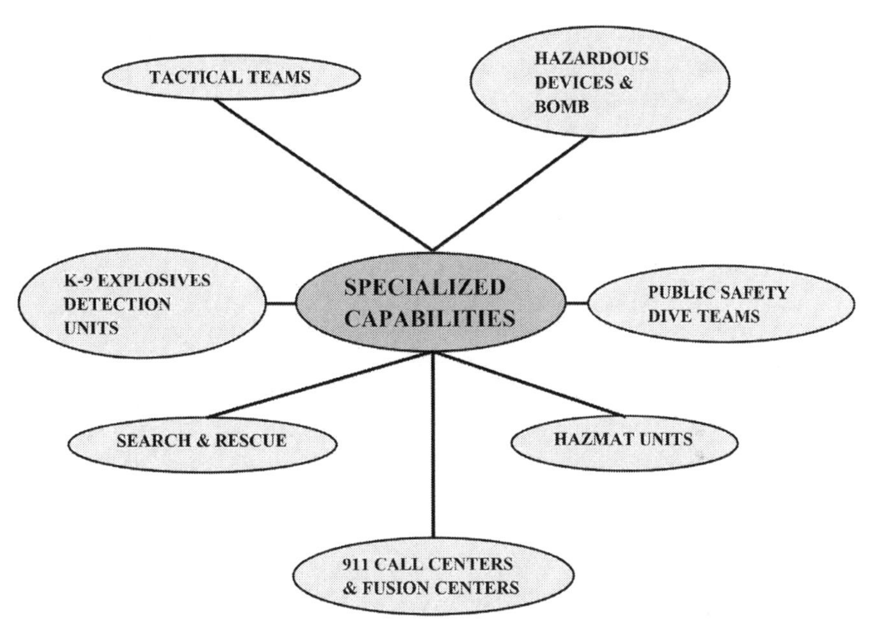

Figure 1.4. Emergency Services Sector Specialized Capabilities

smallpox, and tuberculosis; chemical hazard emergencies; and cyberattacks. ESS personnel perpetually face the challenge of sustaining response levels to existing threats, together with responding to evolving threats requiring new or expanded ESS capabilities (FEMA, 2019).

MAJOR CROSS-SECTOR INTERDEPENDENCIES

ESS operations may be interdependent or closely integrated, in some way or other, with all sixteen critical infrastructure sectors. This close integration creates interdependencies that can cause a disruption in one sector to affect operations in another. Local disruptions can cascade (initiate a chain reaction) into multiple jurisdictions, which can then cause a disruption in one sector to affect operations in another sector across a larger geographic area. A major factor contributing to the overall disastrous effect of a terrorist attack against the ESS is the limited visibility (the hidden risks) of the growing threats faced by interdependent sectors—spillover risks from other sectors that ESS owners and operators can't adequately anticipate or manage.

With regard to the interdependencies of critical infrastructure sectors, the lifeline functions—energy, water, communications, and transportation systems—are those that are essential to the operation of most critical infrastructure sectors and the surrounding communities. Identifying lifeline functions, especially those that are interdependent with other sectors, can help owners and operators prepare for and mitigate the loss of these services in an emergency. In the ESS, transportation systems provide lifeline functions and are also essential to the saving of lives component.

THE BOTTOM LINE

In addition to the critical dependencies of the emergency services sector mentioned to this point, it is important to note that the physical security of ESS facilities (covered later) is important, but actually the focus of the security of this sector is a different kind. Much of the ESS is vulnerable to cyberattack from either inside or outside the network. Cybersecurity is addressed later. Also, ESS infrastructure is extremely dependent on the information technology sector (covered in detail in another volume in this series).

The bottom line is this: any number of interdependencies could threaten one of the key components of our critical national infrastructure—the ESS. Without the active participation of each of these sectors, the key and essential facilities of the ESS could be very vulnerable, a fact that is not acceptable for the protection of our national interests (DHS, 2019).

REFERENCES AND RECOMMENDED READING

BLS. (2018). "Industries at a Glance." Washington, DC: Bureau of Labor Statistics. https://www.bls.gov/iag/tgs/iag332.htm (also 333, 334, 335, and 336), accessed May 17, 2023.

DHS. (2019). "Emergency Services Sector." https://www.dhs.gov/cisa/emergency-services-sector, accessed May 17, 2023.

FCC. (2017). "Public Safety Tech Topic #19—Communications Interdependencies."

FEMA. (2019). "About the U.S. Fire Administration." https://www.usfa.fema.gov/about, accessed May 17, 2023.

Haimes, Y. Y. (2004). *Risk Modeling, Assessment, and Management*. 2nd ed. New York: Wiley.

Henry, K. (2002). "New Face of Security." *Government Security* (April): 30–37.

NAM. (2018). "Facts about Manufacturing." https://www.nam.org/facts-about-manufacturing, accessed May 17, 2023.

NSHS. (2006). *National Strategy for Homeland Security*. https://www.dhs.gov/national-strategy-homeland-security-october-2007, accessed May 17, 2023.

Sauter, M. A., & Carafano, J. J. (2005). *Homeland Security: A Complete Guide to Understanding, Preventing, and Surviving Terrorism*. New York: McGraw-Hill.

Spellman, F. R. (1997). *A Guide to Compliance for Process Safety Management/Risk Management Planning (PSM/RMP)*. Lancaster, PA: Technomic Publishing.

Chapter 2

Emergency Services Sector Risks and Goals

RISK MANAGEMENT PLANNING

The emergency services sector comprises an estimated 4.6 million career and volunteer professionals within five primary disciplines: law enforcement, fire and rescue services, emergency medical services, emergency management, and public works. Encompassing a wide range of emergency response functions, the ESS saves lives, protects property and the environment, assists communities impacted by disasters, and aids recovery from emergencies. ESS operations and functions support each of the other fifteen critical infrastructure sectors and assist in maintaining public safety, security, and confidence in the government.

Multiple factors may affect the critical infrastructure security and resilience posture of the ESS. These factors, which influence the current operating environment and associated decision-making processes, stem from environmental, technological, human, and physical causes. As the ESS focuses on protecting other sectors and the public, unique challenges arise in addressing the security and resilience of the ESS as critical infrastructure. The incapacitation of any of the assets, networks, or systems in this sector, whether physical or virtual, could cause significant harm or loss of life, public health threats, or long-term economic loss.

The following are five major focus areas for ESS security and resilience risk management planning consideration.

- Operational Hazards: Serving and protecting the public routinely exposes emergency services (ES) personnel to the adverse conditions (for example, threats of violence or hazardous materials) affecting the communities they serve. This role requires that ES personnel maintain a high level of threat

awareness and associated response capabilities, as well as the capacity to respond to an increasing number of complex challenges. ES personnel perpetually face the challenge of sustaining response levels to existing threats, together with responding to evolving threats requiring new or expanded sector capabilities.
- Cybersecurity: The ESS is subject to a wide range of risks stemming from cyber threats and hazards. Sophisticated cyber-threat actors and nation-states exploit opportunities to steal information and disrupt or threaten the delivery of essential services. The influx of new mobile technologies that enhance ES organizations' capabilities have altered the vulnerability and threat landscape affecting the sector.
- Natural Hazards: Natural hazards—adverse events caused by the earth's natural processes, such as floods, tropical cyclones, wildfires, tornadoes, earthquakes, and tsunamis—have the potential to cause substantial loss of life and property damage. Responding to disasters caused by natural hazards is one of the ESS's primary functions, and the sector maintains robust response capabilities for all types of emergencies. However, the increasing incidence and severity of natural hazards can exacerbate existing challenges to effective response.
- Criminal Activities and Terrorism: Criminal activities and terrorism—the use of violence and intimidation in the pursuit of personal or political aims—can affect the ESS's ability to prepare for and manage emergencies, enforce the law, and provide public safety. Security and resilience issues that hinder the ability of ES personnel to provide essential services include attacks on personnel and assets, theft of vehicles and equipment, and malicious use of unmanned aircraft systems.
- Crosscutting Issues: Issues stemming from infrastructure, social, technological, and economic changes have the potential to disrupt ES, overburden ES organizations, and increase capital expenditures. For the ESS, crosscutting security and resilience issues include managing access and reentry challenges near an incident or event, aging infrastructure, changing populations, and interdependencies with other sectors.

Operational Hazards

Emergency services personnel serve as the nation's first line of defense in preventing and mitigating the effects of physical and cyber threats, whether natural or man-made. In serving and protecting the public, ES personnel may be exposed to a variety of harmful situations, for example, potential exposure to toxic chemicals when responding to a hazardous material spill or physical abuse when attempting to render aid to an injured person. Effects may be acute and immediately felt but may also accumulate over time, potentially degrading physical and mental health. Security and resilience issues relating to operational hazards include the threat of exposure to hazardous materials, operational burnout associated with elevated and prolonged stress, and the potential for encountering violence and assaults while performing duties.

Exposure to Hazardous Materials

When responding to incidents, ES personnel may face the same hazards affecting the individuals they are attempting to assist. The presence of chemical, biological, radiological, nuclear, or explosive (CBRNE) substances, incomplete or inaccurate cargo information, or exposure to other unknown substances can create a dangerous response environment for ES personnel.

- CBRNE: CBRNE substances represent a broad range of potential threats that may be present at an incident, and identifying the type and properties of these substances requires training. Exposure can have serious long-term consequences. Among EMS workers, exposure to harmful substances such as potentially infectious fluids (e.g., bodily fluids) was the second leading cause of occupational injuries after body motion injuries.[1] Similarly, exposure to powerful synthetic opioids such as fentanyl and carfentanil, especially when encountering illicit production or milling operations, endangers the safety of responders and requires new approaches to maintain safety.
- Incomplete/Inaccurate Cargo Information: ES personnel rely on cargo information to help inform their response, and inaccuracies or incomplete information can lead to unnecessary exposures. Personnel must rely on sensors on either their persons or their vehicles to identify the presence of unexpected dangerous substances.
- Unknown Substances: ES personnel can be exposed to unknown substances intended for others, for example, when screening packages for hazardous materials. Similarly, ES personnel may respond to an incident and discover on-site, or after the fact, that hazardous materials were present.

Operational Burnout

ES personnel have demanding roles and responsibilities. In many cases, the daily duties of ES personnel require interacting with and providing aid to people on their worst day. The stress of providing emergency response coupled with high operational tempos, workplace constraints, training and certification requirements, and the pressures of daily life can lead to operational burnout. Both personnel and communities may suffer from the results of operational burnout. Public expectations have increased despite limited funding and resources. These constraints encumber ES organizations' ability to retain experienced personnel and recruit new personnel, which can contribute to an overall reduction in available essential services.

- Operational Tempo: ES personnel often work extended shifts of ten hours or more. A recent study of a sample of EMS personnel found that the risk of injury and illness increased with shift length.[2] A similar study of firefighters found that additional twenty-four-hour shifts—beyond a standard work schedule of between eight and eleven twenty-four-hour shifts per month—and increased job demands were associated with elevated blood pressure.[3] Heart attacks and strokes caused the majority of on-duty deaths in 2017.[4] In addition, extended periods of heightened awareness in anticipation of emergencies happening throughout a shift may cause high levels of stress. ES personnel are frequently less likely to ask for help than those in other professions, allowing stressors to go undetected and unaddressed.[5]
- Increased Public Expectations: The public increasingly expects first responders to possess the capabilities to respond to emergencies of all kinds. For example, firefighting has become a smaller portion of the specialized services fire departments provide to their local communities, giving way to calls for EMS or other specialized services. Trying to meet the expanded demand stresses ES organizational budgets, increases personnel training requirements, and raises costs associated with equipment maintenance.
- Funding and Resource Constraints: Diminished government budgets may affect the capacity of the ESS to adequately address, anticipate, or prepare for changes in the sector's risk profile. As costs increase for health care, personnel, fuel, and equipment maintenance, along with the demand for essential services, ES organizations may find it difficult to rapidly respond to incidents, maintain the required capabilities, train specialized teams, or replace aging equipment necessary to support their communities.
- Workforce Shortages: Workforce shortages in any of the ESS disciplines may hinder their ability to respond effectively and efficiently. When ES organizations are challenged to perform a greater amount of work with fewer personnel, this situation can cause more rapid burnout as well as increased spending (e.g., for overtime and equipment maintenance).

Violence and Assaults

Potential violence and assaults against first responders are major security and resilience issues, which can manifest as physical assaults as well as verbal threats and abuse. In performance of their duties, ES personnel routinely interact with unruly individuals. EMS personnel face a growing number of violent incidents (e.g., physical assaults, verbal threats and abuse, and intimidation), which nearly doubled from approximately 1,800 incidents in 2009 to 3,500 in 2016.[6] Law-enforcement officers are three times more likely to sustain injury than all other workers, and assault-related injuries to law-enforcement officers grew nearly 10 percent annually between 2003 and 2011 while rates of injuries in all other professions remained unchanged, according to data from the National Institute of Occupational Safety and Health.[7]

- Physical Assaults and Related Injuries: While training helps all first responders address such threats, assaults can nonetheless result in serious injury to first responders. The Centers for Disease Control estimates that 3,500 EMS workers sought treatment at an emergency department in 2016 because of violence.[8] Because ES personnel generally receive training to reduce the chance of personal injury during an emergency, some may blame themselves for becoming the victim of an attack, which can negatively affect their mental health. In addition, physical assaults and concerns about the safety of first responders can have negative impacts on retention and recruitment, hurting the organization.
- Verbal Threats and Abuse: First responders may face verbal threats and abuse from patients, individuals being detained, or bystanders. Reasons can range from anger over the timeliness of ambulance arrival to frustration arising from a sense of helplessness in the situation. Verbal abuse adds stress to an already stressful situation and over time can contribute to larger health issues for ES personnel.

Case Study: Protecting the Protectors

Numerous studies have found that a significant percentage of first responders to the World Trade Center attacks on 9/11 have had posttraumatic stress disorder (PTSD), as well as associated cognitive difficulties and major depression (for example, PTSD is reported at anywhere from 7 to 24 percent, depending on the first responders studied and the number of years since the event). Medical team workers responding to the 2011 Tōhoku, Japan, earthquake and tsunami are reported to have suffered from clinical depression at a rate of over 20 percent. Two police officers who responded to the Pulse nightclub shooting in Orlando, Florida, have spoken publicly about the resulting PTSD; one, deemed disabled, was ultimately relieved of his duties.

Struggling in the wake of significant—perhaps historical—tragedies is perhaps unsurprising, but first responders face challenging situations as a matter of course. Their duties lead them into high-stress—and often high-risk—scenarios, often in close succession. Stressors include repeated exposure to death, grief, and injury, as well as threats to personal safety. With limited time to process these experiences, our front line may experience PTSD, depression, suicidal ideation, and a host of related conditions. Worse still, many of those who serve may refuse to seek help for various reasons: some feel they should be able to handle the stress on their own, some recognize a stigma attached to behavioral health conditions, and some lack the necessary time and financial resources.

Fortunately, a cultural shift has taken place, emphasizing awareness and support programs to combat these occupational hazards. ES leaders have implemented steps to safeguard the physical and mental health of their personnel. New staff should be carefully screened for suitability to handle extreme stress, and all staff may benefit from mental health training. Providing planning, training, and clear roles and reporting structures in advance of an event provides comfort through a sense of preparedness. During an event, leaders should continually assess the team's welfare, and team members can employ a "buddy system" for both stress checks and physical safety. Follow-up procedures for particularly difficult experiences should include counseling and debriefing for responders, as well as time away from stressful assignments when possible.

Cybersecurity

The ESS is subject to a wide range of issues stemming from cyber threats and hazards. Sophisticated cyber actors and nation-states exploit opportunities to steal information and disrupt or threaten the delivery of essential services. The influx of new mobile technologies that enhance ES organizations' capabilities may also alter the vulnerability and threat landscape affecting the sector.

Issues of higher cybersecurity risk for the ESS include advanced persistent threat (APT) attacks, distributed denial-of-service (DDoS) attacks, increased connectivity and disruptive digital technology, and malware and ransomware. Recognizing and mitigating these issues could help to limit cyber intrusions.

Advanced Persistent Threat

Coordinated campaigns by motivated cyber-threat actors pose significant risk, and opportunities will likely continue to be found for attacks on cyber assets. APTs can exploit these opportunities to establish persistence in a network and acquire sufficient access to achieve objectives (e.g., exfiltration of sensitive information), given enough time and resources. An ES organization's information systems could be compromised by an APT, hindering response efforts.

VPNFilter and Dragonfly represent recent prominent examples of malware and a cyber-threat actor that could affect the sector.

- VPNFilter: In May 2018, Cicso's Talos Intelligence Group announced its research into a modular malware system they named VPNFilter, which had infected more than five hundred thousand devices. The malware uses vulnerabilities in a range of network devices—primarily internet routers—to install a persistent foothold in the targeted devices, which can be used to deploy further modular malware on those devices.[9]
- Dragonfly: Russian government cyber-threat actors have been targeting U.S. critical infrastructure sectors since at least March 2016 in a coordinated campaign of malware attacks collectively named Dragonfly. The threat actors used a combination of spear phishing (highly targeted emails with malicious attachments) and watering-hole attacks (introducing malware through well-known industry trade publications' websites) to collect user credentials. The threat actors were able to establish footholds in the target networks and conduct network reconnaissance, move laterally, and collect sensitive or proprietary information.

Distributed Denial-of-Service Attacks

DDoS attacks are a growing threat, generating immense bandwidth loads to the point of disruption or creating openings for malware to be deployed. As ES organizations introduce more internet-connected devices (see "Increased Connectivity and Disruptive Digital Technology" below) into their operations, more vulnerabilities to DDoS will arise. Devices used in response activities that are connected to the internet could have their functionality diminished by DDoS, rendering response activities less effective. Botnets have been used to leverage internet-connected devices to carry out DDoS attacks, and techniques such as amplification potentially extend the potential for disruption. The sector's reliance on telephony, especially for public services answering points (PSAPs), potentially exposes ES organizations to telephonic denial-of-service attacks.

- Botnets: Botnets are collections of internet-connected devices that have been infected with malware to respond to specific requests from a command-and-control entity. Potential devices range from home computers to internet of things (IoT) devices. Botnets can be used to generate massive amounts of internet traffic to a specific target with the intention of disrupting essential services. A recent high-profile example was the Mirai botnet, which was used in October 2016 in a DDoS attack on a major domain name system

(DNS) service provider. The attack flooded 1.2 terabits per second (Tbps) of internet traffic (at the time, the highest volume of DDoS traffic ever recorded) managed by the DNS provider and shut down many well-known websites. At the height of the attack, millions of users were denied internet services in North America and Europe. Similar to the previous September 2016 Mirai attack, the DNS attack employed millions of compromised internet-connected security cameras to simultaneously conduct the attack.[10]
- Amplification: Amplification refers to a technique in which a cyber-threat actor abuses internet-connected devices such that they respond to a small packet of code from the attacker by sending large packets of data to a target as part of a DDoS attack. The effect amplifies the bandwidth sent by the threat actor, resulting in much larger amounts of data flooding the target. Unlike with botnets, the threat actor does not necessarily need control of the device. Instead, a threat actor abuses the devices' intended functionality to respond to requests and causes the responses to flood a target's servers. Memcached DDoS attacks, a specific type of amplification attack, resulted in 1.3 Tbps and 1.7 Tbps of internet traffic in separate attacks in March 2018, though no critical services were disrupted.[11]
- Telephonic Denial of Service (TDoS): Cyber-threat actors can leverage malicious code to flood public services answering points with fraudulent calls and disrupt operations. Malicious code can abuse mobile phones' functionality not only to place fraudulent calls to 911 without the users' knowledge or consent, but also to propagate the malicious code to other users and create more calls.[12]

Increased Connectivity and Disruptive Digital Technology

Computer-aided dispatch (CAD) improves the capabilities of operators. Location services, fleet management, environmental sensors on responders, and other digital technology improvements aid responders with increased connectivity. This increased connectivity coupled with disruptive digital technology—new technology that displaces an established technology or means of operation—may also create new risks if the systems are not properly secured. In general, combining physical and digital technologies may introduce new risks, including increased points of access through which malicious code could be introduced or data could be stolen, and the potential for cascading failures due to interconnectivity.

- Wearables: With the forthcoming Nationwide Public Safety Broadband Network, ES personnel could expand the use of wearable communications devices and sensors that can provide beneficial functions such as authentication, heart-rate monitoring, video recording, hands-free communication, or location tracking. These new and expanded capabilities represent greater security needs for data and communications.

- Next Generation 911: Next Generation 911 systems operate using digital (rather than analog) technologies. The use of modern digital networking offers many benefits that enhance PSAP capabilities, but it also requires administrators to manage associated cyber threats and vulnerabilities.
- Increased Points of Access: An expanding footprint of networked devices introduces more points of potential targets for cyberattack in the network. Both physical (e.g., locations for input or display devices) and cyber (e.g., network ports) points of access could be exploited.
- Cloud Services: ES organizations are increasingly incorporating cloud services into their operations. Cloud software-as-a-service (SaaS) is leveraged to enhance response capabilities. Although cloud services offer benefits, such as high availability, advanced data analysis and storage, and decreased ownership cost, new cybersecurity concerns are associated with those benefits. Cloud services share many of the same cybersecurity issues as physical, on-site information technology (IT) (e.g., denial of service, APT, stolen credentials, and phishing) yet also exhibit virtual susceptibility to attacks, including malicious control of virtual machines and attacks on systems running virtual processes.
- Cascading Failures: Automated systems that are dependent on interconnected devices may be subject to cascading failures that result from disruptions along the network of devices. A disruption within a chain of interconnected devices can have drastic cascading effects on the safety of ES personnel or the availability of supporting capabilities of ES personnel.

Malware and Ransomware

Malware, including ransomware, is commonly used in attacks on many types of ES organizations' networks. Malware could be used to steal an organization's information, to steal data such as personally identifiable information or medical information, or to interrupt business operations. CryptoWall, Emotet, and WannaCry are three prominent recent examples that have affected several critical infrastructure sectors.

- CryptoWall: CryptoWall is among the most commonly used ransomware varieties, with various forms of the ransomware targeting hundreds of thousands of individuals and businesses. The ransomware arrives on the affected computer through spam emails. Not only does it encrypt files and prompt the business to pay for the key, but it also hides inside the operating system and adds itself to the Startup folder, accesses passwords, and deletes volume shadow copies of files so that data restoration is difficult or impossible. Some reports estimate that CryptoWall has grossed over $325 million in ransom payments since 2014.[13]

- Emotet: Predominantly targeting the financial sector but affecting municipalities as well, Emotet represents an especially complex type of malware because it is advanced, modular, and polymorphic (i.e., it changes its own content to evade signature detection). Potential consequences include temporary or permanent loss of sensitive or proprietary information, disruption to regular operations, financial losses incurred to restore systems and files, and potential harm to an organization's reputation.
- WannaCry: In May 2017, WannaCry ransomware infected organizations all over the world, encrypting and paralyzing the systems. WannaCry exploited security vulnerabilities in Windows computer systems. Information about the Windows code flaw was released in leaked National Security Agency documents, and although a patch was developed to eliminate the security vulnerability, many organizations did not download the upgrade before the attack. The ransomware affected mostly business control systems, although the malware mechanism used could be adapted to disrupt process control systems.

Supply Chain Cybersecurity

ESS assets and networks are susceptible to compromised vendor communications associated with the sector's supply chain. Email phishing attempts from presumed trusted vendor email accounts are becoming more frequent. Successful phishing attempts could allow cyber-threat actors remote access to enterprise networks and the opportunity to escalate attacks to operations infrastructure. Trusted contractors and vendors may have legitimate remote access to provide services; however, this access could turn problematic if the contractor or vendor has been compromised. The supply chain for software itself represents another cybersecurity issue, as compromised software introduced along the supply chain could be used to attack ESS networks.

- Third-Party Attacks: Cyber-threat actors have targeted critical infrastructure subcontractors' networks to abuse access the subcontractor might have to the target organization. This abuse of trust in software suppliers and subcontractors can affect even well-protected organizations.
- Software Supply Chain: In 2017, software supply-chain attacks increased dramatically across all sectors.[14] In attacking software providers, cyber-threat actors replace legitimate business software with maliciously modified versions, unbeknownst to end users. If an ESS organization attempts

to download, for example, the latest version of previously trusted software, it receives the malicious version instead.

Case Study: Thwarting TDoS Attacks

In October 2016, an eighteen-year-old hacker played a digital "prank" that ultimately targeted multiple 911 departments in Arizona, California, and Texas. He tweeted a link to a web page that, when visited on a mobile phone, would cause the phone to place repeated 911 calls. The resulting volume of calls to 911 operators threatened availability of services. Such a strategy is perhaps clever but not overly complicated to enact, yet it is difficult to foresee and prevent. As people place more emergency calls from cell phones, the potential for TDoS attacks to resemble DDoS attacks rises.

Recent years have seen an exponential increase in the frequency and intensity of DDoS events. In response to this escalating threat, the U.S. Department of Homeland Security (DHS) is working on a multimillion-dollar effort to protect critical digital systems from DDoS attacks. DHS has provided $14 million in grants for DDoS-prevention studies, including preventing TDoS attacks. Results to date include a prototype technology that can detect and thwart fake telephone calls and the selection of multiple partners to pilot the technology. However, there are concerns about blocking any calls, even likely false ones, to 911 centers, so the technology must be made infallible or another solution must be sought.

Natural Hazards

Natural hazards include major adverse events caused by the earth's natural processes, including floods, cyclones, wildfires, tornadoes, earthquakes, and tsunamis. Natural hazards can cause disasters that result in loss of life and property damage, as well as economic damage and disruption or destruction of facilities and operations. The severity of a disaster is measured in terms of lives lost, economic disruption, and the affected population's ability to rebuild.

Responding to disasters caused by natural hazards is one of the ESS's primary functions, and the sector maintains robust preparedness and response capabilities for all types of such hazards. Major issues of natural hazards for the sector include larger and more severe hazards and the environmental dangers that arise because of them.

Cost, Frequency, and Severity of Natural Disasters

Recent billion-dollar-loss natural disaster events in the United States include Atlantic and Gulf Coast hurricanes, northeastern winter storms, freezing in southeastern states, tornadoes and hailstorms in central states, and wildfire and drought in western states. In 2018, the United States experienced fourteen billion-dollar disasters, with total damage costs across all sectors exceeding $91 billion. The total number of these disasters is the fourth-highest behind 2011, 2017, and 2018.[15] Such large-scale events have cascading impacts across sectors and regions. More extreme weather increases the geographic magnitude and severity of events caused by natural hazards, requiring a surge of ES resources, often for extended periods, while also straining resources in partnering regions that might otherwise supply mutual aid. Such events cause an increased hazard to responders while often disrupting critical services needed for effective response.

- Emergency Management: Disaster impacts are increasing. Large-scale disasters such as Hurricane Harvey can make management difficult from the outset. Wide-ranging damage can affect primary and secondary locations for emergency operations centers, requiring makeshift workarounds and delaying response efforts. High winds, wildfires, ice storms, and other hazards often damage communications infrastructure, compounding the challenges of managing increased communications needs (see the "Cross-cutting Issues" section below).
- Public Works: Major natural hazards typically cut off or drastically restrict access to affected areas. This hinders response efforts and may initially curtail debris removal operations, damage assessments, and restoration efforts.
- Police, Fire, and EMS: Severe natural hazard events can have impacts on primary operating and support facilities (e.g., local police stations, firehouses, and storage facilities) and stress organizational preparedness efforts. Primary facilities may be damaged by fire or flooding or may experience a prolonged power disruption, reducing response capabilities. Prolonged emergencies or preexisting conditions caused by extreme weather patterns (e.g., droughts, floods, and wildfires) may complicate response efforts, strain limited resources, or increase operational burnout.

Environmental Dangers

ES personnel may be exposed to health risks associated with natural disasters, including toxins dispersed from flooding (e.g., sewer and wastewater), carcinogens as by-products of fires, or infectious diseases. Also of concern during ESS operations are physical hazards caused by natural disasters, such as falling or flowing debris, building or roadway instability, or extreme heat or cold.

- Release of Toxins: Flooding can cause release and dispersion of unknown toxins, which creates a health hazard for ES personnel. Floodwaters can contain human and livestock waste; household, medical, and industrial hazardous waste; coal ash (which contains arsenic, chromium, and mercury); or other contaminants that cause adverse health effects.
- Carcinogens as By-Products of Fires: Exposure to wildland smoke, even at low to moderate levels, represents a safety and health hazard to wildland fire personnel. Wildland fire smoke contains a variety of inhalation irritants, including carbon monoxide, aldehydes, particulate matter, crystalline silica, and polycyclic aromatic hydrocarbons. Some of the compounds in wildland fire smoke are confirmed carcinogens or suspected carcinogens.[16]
- Infectious Diseases: Pandemics—a type of natural hazard themselves—affect ES organizations, as some ES personnel, or members of their families, will likely fall ill during a pandemic, reducing workforce availability just as demand for emergency services increases. Conditions in the wake of hurricanes and other natural hazards can increase the probability of the spread of infectious diseases.
- Physical Hazards: While public works personnel will clear, remove, and dispose of debris from natural hazards, ES personnel operating in time-sensitive situations need to take extra precautions to avoid physical hazards, which can impede their operations.

Case Study: Responding to Flooding and Fires at a Chemical Plant

Hurricane Harvey led to explosions at a major U.S. chemical facility in Texas. The U.S. Chemical Safety Board (CSB) investigated the event and found that the facility had established—and did follow—policies and safeguards for hurricanes. These included elevating portable equipment to keep it out of floodwater, staging sandbags, acquiring a boat and forklift that could operate in floodwater, and activating a "ride-out crew" who would remain on-site during the storm. Workers also moved organic peroxides from low-temperature warehouses, where the peroxides were normally stored, to nine refrigerated trailers used for shipping. Six of the trailers were relocated to a high-elevation area, but three could not be moved and were in danger of losing refrigeration. When the storm's devastating potential became evident, plant officials determined that the refrigerated trailers might very well lose power, causing the organic peroxide products inside to combust within a few days. Plant personnel alerted local emergency responders, who evacuated the ride-out crew and then implemented an evacuation zone around the site.

Concurrent emergency response remained underway throughout the region to address the havoc in Harvey's wake. To facilitate transport of rescue resources through the area, local officials kept the main highway open, although it ran through the evacuation zone. Five police officers drove on this route through a cloud of smoke, later confirmed as coming from the chemical manufacturing facility, and shortly thereafter began to have nausea, headaches, sore throats, and watering eyes. Officials then shut down the highway, but a total of twenty-one people sought medical attention from reported exposure to the noxious fumes. Within twenty-four hours of the road closure, the chemicals in all three at-risk trailers caught fire.

The status of the chemicals in the six remaining trailers was unclear. On September 3, eight days after Harvey made landfall in Texas, emergency responders entered the site and conducted a controlled burn of the remaining trailers to end the evacuation and allow citizens to return to their homes.

Criminal Activities and Terrorism

Criminal activities and terrorism affecting critical infrastructure make headlines around the world almost every day. Terrorism, which can be described as the use of violence and intimidation in the pursuit of ideological aims, can take many forms, including chemical, biological, radiological, nuclear, and explosive attacks.

In the ESS, security and resilience issues regarding criminal activities and terrorism include direct and secondary attacks as part of violent extremism, theft of vehicles and equipment, and malicious use of unmanned aircraft systems (UASs). Recognizing and mitigating these risks could help to limit the financial, operational, and human impacts of criminal activities and terrorism.

Homicides, Assaults, and Targeted Attacks

Every year, a substantial number of ES personnel are killed or injured in the line of duty. Among all occupations, not just the ESS, police and sheriff's patrol officers count among the highest numbers of workplace homicides. Similarly, ES personnel, not just law enforcement, face some of the highest rates of workplace violence in the course of performing their duties. In addition, terrorists and violent extremists have attacked ES personnel, especially law enforcement, in recent years, and terrorist organizations continue to encourage supporters to carry out both direct and secondary attacks on first responders.

- Felonious Homicides and Assaults: Between 2014 and 2018, 259 law-enforcement officers died in the line of duty as a result of felonious incidents. Of these, 109 were killed as part of investigative or enforcement actions, and 53 were killed in ambush-style attacks.[17] In addition, more than 60,000 officers were assaulted in 2017 alone, with more than 17,000 attacks resulting in injuries.

- Attacks Related to Terrorism and Violent Extremism: In 2016 and 2017, fifteen attacks targeting law enforcement resulted in twelve fatalities and twenty-two injuries, according to the University of Maryland's Global Terrorism Database.[18] Of the fifteen attacks, six were perpetrated by people with expressed antigovernment or antipolice sentiments; another four perpetrators had unknown motivations but nevertheless targeted police or police headquarters. Most attacks involved armed assault on law enforcement, typically shootings of officers or their cruisers. Several other attacks involved incendiary devices targeted at police headquarters.

Theft and Impersonation

ES equipment can be a target for theft; objects of particular interest include response vehicles, uniforms, firearms, access control items (e.g., badges or keys), supplies (e.g., medicine or narcotics), and technological equipment. When an unauthorized person gains access to emergency response vehicles or equipment, the acquisition could lead to a dangerous situation. The theft of emergency response vehicles and equipment may be an indicator of preoperational activity by a malicious actor or actors that constitutes a potential threat to public safety, as stolen emergency vehicles and equipment have been used to exploit site vulnerabilities, destroy critical infrastructure, and harm people and property.

- Stolen Vehicles and Equipment: In the course of responding to an emergency, ES personnel may inadvertently present an opportunity to steal an emergency response vehicle. Focused on the immediate needs of injured people, ES personnel may not fully secure a vehicle when arriving at a scene. Theft of ES equipment not only affects the safety of emergency responders but also jeopardizes their ability to provide timely and effective public safety services.
- Impersonation: Both international and domestic malicious actors may exploit the public trust in first responders by impersonating first responders to inflict harm, exploit site vulnerabilities, or destroy critical infrastructure. Malicious actions include the acquisition of authentic or fraudulent uniforms, equipment, vehicles, and other items that may be associated with law-enforcement, fire, and EMS personnel.

Unmanned Aircraft Systems

UASs offer both promise and peril. They have been used to survey areas affected by natural disasters, such as the city of Houston following Hurricane Harvey, to effect a more informed response.[19] While official use can improve response efforts, nonofficial use is on the rise, as UASs are commercially available and easy for the layman to use. The upward trend suggests that negative encounters with UASs, too, will continue to increase in the near future. UASs could be used for illicit surveillance to target or monitor ES personnel or response operations, to disrupt operations (e.g., interfering with emergency operations or initiating cyberattacks that compromise ES communications and networks), or to cause physical damage.[20]

- Surveillance: UAS video capabilities can be used by adversaries for preoperational planning to monitor and assess security operations at sensitive sites, large-scale events, and law-enforcement or emergency response operations.
- Disruption: Intentionally or unintentionally, UASs may endanger law-enforcement, medevac, firefighter, and other ES flight operations by operating in the same airspace. Adversaries can also direct UASs close to a facility to access, monitor, or attack computer networks and/or to monitor or interfere with radio frequency communications. Such UAS proximity to a facility, whether intentional or unintentional, can harass, hinder, or inhibit emergency response operations.
- Weaponization: UASs can be central to an attack intended to cause casualties or physical damage; possible strategies include disrupting air traffic, deliberately crashing, and delivering a hazardous payload (e.g., an explosive device or a chemical, biological, or radiological weapon).

Case Study: Fighting on Two Fronts

From a safe distance, wildfires make for spectacular viewing, as do aerial firefighting teams' efforts to battle the blazes. It is perhaps unsurprising, then, that UAS hobbyists send in their drones to get prime footage. Unfortunately, the response teams find the tiny fliers a nuisance at best, and a danger at worst. On at least twenty-two occasions during the 2018 wildfires, fire aircraft were forced to halt proceedings temporarily when drones were discovered flying close enough to risk collision.

Aerial firefighting requires speed and rapid directional changes, as fliers chase hot spots, swoop downward to release the payload, and quickly fly away. Much of the flying takes place fairly low to the ground, in the same airspace occupied by hobbyist drones. Unauthorized UASs have the potential to distract pilots or crash into planes. Aircraft are somewhat unstable just after release, increasing the risk factor of a UAS encounter. Hobby drones not only threaten the safety of the firefighters—both those in the plane and those beneath them—but also delay the firefighting operations, imperiling citizens and their property.

The increasing frequency of UAS appearances over or near wildfires has occasioned multiple public outreach efforts, including a tagline shared across fire agencies: "If you fly, we can't." Other public tools include the U.S. Department of the Interior's wildfire location data-sharing program, "Current Wildland Fires," and the Federal Aviation Administration's smartphone app B4UFLY.

Crosscutting Issues

The ESS is subject to several crosscutting issues that stem from infrastructure, social, technological, and economic changes. Crosscutting security and resilience issues include access and reentry challenges, aging infrastructure, changing populations, communications challenges such as interoperability, and dependencies and interdependencies with other sectors. These issues could disrupt ES operations, overburden ES organizations, and increase capital expenditures.

Access and Reentry

Effective control and coordination of access for key response and recovery resources into an affected area before, during, and after an emergency improves the likelihood of successful community recovery. The process of managing access into restricted areas or through emergency zones during an incident is a state or local responsibility and can become increasingly difficult when disasters extend across multiple jurisdictions or involve significant population evacuations.

- Multi-jurisdiction: When an incident extends across jurisdictional boundaries, the number of stakeholders requiring access to conduct damage assessments, protect critical infrastructure, and reestablish essential services increases, adding complexity to response and recovery efforts. A lack of a common approach to access management may hamper restoration efforts, increase recovery costs, and reduce overall operational success.
- Evacuation: During incidents that require significant population evacuations, access management is particularly important to ensure coordination of public- and private-sector response and recovery assets, restoration of

critical infrastructure and essential public services, and the safe and orderly return of community members back into an affected area. A lack of coordinated access management can cause confusion among stakeholders, negatively affect restoration efforts, and create hardships for both residents and first responders.

Aging Infrastructure

A significant portion of U.S. infrastructure is in need of repair or replacement. The American Society of Civil Engineers' 2017 Infrastructure Report Card rated U.S. infrastructure as a whole at D+. Of that, roads received a D, bridges a C+, ports a C+, rail a B, inland waterways a D, energy a D+, drinking water a D, and wastewater a D+. The state of disrepair and viability of our nation's infrastructure is of great concern to the ESS, as all ES disciplines rely on this infrastructure (e.g., functioning electrical grids, water management systems, roads, and bridges) to perform critical functions. This section highlights issues of concern for these sectors.[21]

- Roads: The nation's roads and highways are commonly overcrowded, in disrepair, and significantly underfunded. In 2014, over $160 billion was wasted in time and fuel owing to traffic delays and congestion. Approximately 20 percent of highways are in poor condition, causing increased costs of vehicle maintenance and repairs. An approximate backlog of over $700 billion in projects awaits funding to repair existing highways, make strategic expansions, and update the highway system (e.g., for safety, operational, and environmental improvements).
- Bridges: In the United States, most highway bridges are designed for a life span of approximately fifty years. Of the more than six hundred thousand bridges in the United States, approximately 40 percent are fifty years old or older, and 9 percent are structurally deficient. Although bridge conditions have improved in recent years, funding for bridges may be inadequate to maintain or improve current capacities. An estimated $123 billion is needed to eliminate the nation's bridge upgrade backlog.
- Ports: The vast majority of the nation's international trade—99 percent—flows through its ports, accounting for approximately 26 percent of its economy. As the ships carrying this cargo continue to increase in size and capacity, U.S. ports become more congested and less able to accommodate the largest ships. Ports are expected to spend approximately $155 billion from 2016 to 2020 to expand, modernize, and repair in response to demands of international trade. Connected infrastructure (land, rail, and

inland waterway connections to ports) requires commensurate aid, yet funding for these improvements and repairs is lacking.

- Rail: The freight rail industry has made important investments and repairs in the past several years to improve its systems and meet future needs. Short rail lines are in need of upgrading and maintenance funding—more so than long-distance lines—to advance in freight car capacity and repair and replace bridges.
- Inland Waterways: A total of fifty thousand miles of canals, locks, and dams comprise the United States' inland waterways system, the majority of which is older than the original fifty-year design life of its components. These waterways are an important part of freight transportation, connecting ocean ports with inland transportation hubs and accounting for approximately 14 percent of domestic freight. Age and disrepair, due to lack of funding, result in frequent delays for hours at a time, contributing to economic losses. Although investments have been increasing in recent years, repair and upgrade projects can take decades to complete.
- Energy: While energy infrastructure owners—predominantly in the private sector—are investing to ensure long-term capacity and sustainability, the energy sector depends on complex systems composed of assets that vary widely in age and condition. Almost half of U.S. natural gas transmission and gathering pipelines were built in the 1950s and 1960s; annual investments in interstate natural gas pipelines will total more than $2.6 billion through 2030.[22] Aging electricity transmission and distribution lines must be replaced or upgraded; spending on distribution systems grew to $51 billion annually as of 2017, and spending on transmission infrastructure rose to $21 billion annually as of 2016.[23]
- Water and Wastewater: Renewal and replacement of aging water and wastewater infrastructure has been the top issue for the water industry between 2014 and 2018.[24] Many of the nation's one million miles of pipes transporting drinking water, which have life spans of seventy-five to one hundred years, were laid in the early to mid-twentieth century. Replacing pipes, pumps, and other assets will require significant investment. Old wastewater conveyance and treatment systems may not meet the demands of growing populations; an estimated $271 billion in investment will be needed to meet current and future demands.

Changing Populations

Increased population density in urban and suburban areas and changing population characteristics may exacerbate existing risks. Dynamics that could alter the resource requirements of ES organizations include growing populations leading to greater population density, greater mobility, and overall aging of the general population.

- Population Density: A growing population may create areas of higher population density, especially in urban centers, which increases response needs. Urban settings also amplify many of the issues surrounding incident management response. The density of the environment and the need to facilitate evacuation, secure the scene, and establish restricted areas may stress emergency response resources.
- Mobility: Global mobility increases the potential for the spread of biological agents and communicable diseases and the related loss of able-bodied ES personnel responding to those incidents. Evolving health issues like antimicrobial resistance could exacerbate these concerns.
- Aging Population: The average age of the general public is increasing (adults over the age of sixty-five composed 15.2 percent of the population in 2016 but are projected to make up more than 20 percent in 2030), which may increase the frequency at which emergency medical response is needed, potentially stressing available resources.

Communications

ESS operations are highly reliant on communications services and equipment to effectively execute mission functions. All manner of ES communications are susceptible to disruptions from natural hazards, cyberattacks, physical attacks, and general coverage and interference issues. With thousands of separate systems, networks, and equipment models, communications interoperability can be a major challenge between different ES organizations or disciplines. Increased adoption of digital and internet technologies for ES communications adds cybersecurity concerns (see increased connectivity and disruptive digital technology in the "Cybersecurity" section above).

- Communication Disruptions: In large-scale response efforts, many additional resources supplement local organizations. Local emergency communications networks may not support the influx of traffic, resulting in busy signals and lack of available frequency. Additionally, the incident itself could affect communications infrastructure, reducing availability.
- Interoperability Issues: In recent years, public safety has seen a rapid expansion in technology advancements and in the type and manner of information sharing among responders and government officials. New applica-

tions and systems have created new challenges for interoperability and for ensuring that the right people receive critical information at the right time.

Dependencies on Other Sectors

Both ES personnel and ES-related physical and cyber assets rely heavily on the resources and continual operation of other critical infrastructure sectors, including the communications, energy, health care and public health, information technology, transportation systems, and water and wastewater systems sectors. In addition, the sector provides essential public-safety-related services to all other sectors, including both support for personnel and restoration of infrastructure on which the other sectors rely.

- Communications: As described above, the ESS relies heavily on operational and public communications, such as through an internal communications network, 911 services, or other public alerting and warning systems.
- Energy: Fuel and electrical power are essential for the sustainment of ESS operations and supporting facilities.
- Health Care and Public Health: In responding to emergencies, EMS and other first responders coordinate with the health-care sector.
- Information Technology: Use of greater automation, CAD, watch and warning systems, and wearable sensors demonstrates ES disciplines' increasing dependence on digital assets and networks to carry out missions. Reliable IT improves services by providing essential support to ES personnel.
- Transportation Systems: To respond to emergencies effectively, the ESS depends on a resilient transportation network. Response vehicles must be able to transport people, goods, and services to and from incident areas. This includes the movement of ES assets to other geographical locations throughout the nation.
- Water and Wastewater Systems: The critical mission of providing emergency services, such as in firefighting and public works, requires a clean and reliable water supply.

Case Study: Transporting Resources in Sandy's Wake

October 2012 brought one of the most destructive hurricanes in recent U.S. history. Hurricane Sandy caused unprecedented storm surges and flooding that devastated communities in twenty-four states, from the eastern seaboard west to Michigan and Wisconsin, as well as nations in the Caribbean, Bermuda, and

Canada. In many areas, the private sector drew on lessons learned from past storms, implementing response and recovery efforts successfully.

Nonetheless, Sandy presented significant challenges, exacerbated in part by transport difficulties, especially for utility crews and other first responders who had to cross state lines to support emergency response efforts. According to the Regional Fleet Movement Coordination Initiative, a one-hour delay in fleet movements can put restoration efforts back an entire day. The region was battling lost power and paralyzed fuel distribution networks, both of which caused cascading difficulties throughout the private sector, highlighting interdependencies between sectors that had previously gone unremarked. Lack of communication and transparency between public and private sectors made coordination difficult, compounding transport and assistance challenges.

Lessons learned in Sandy's aftermath included a clear need to facilitate private-sector fleet and resource movements efficiently while remaining in compliance with state and federal requirements. Recommendations include developing a multistate information database to facilitate fleet vehicle movement, establishing vehicle E-ZPass accounts for emergency fleet response, providing fleet operators with widely accepted documentation in lieu of national credentials, looking to fleets that are successful in moving easily across state lines (e.g., ambulances) for best practices, improving regional communications, and developing standardized rules for response efforts to enable consistency across regions.

EMERGENCY SERVICES SECTOR SECURITY GOALS AND ATTRIBUTES

In responding to the risks and threats outlined above, the Department of Homeland Security has identified eight general attributes and goals of critical infrastructure (including ESS).

- Critical Asset Reduction Goal: Sector resiliency will be most assured if no particular asset can be assessed as more critical than any other. While the ultimate ideal goal would be zero critical ESS assets, the sector will strive to reduce the number of critical assets whenever and wherever possible within fiscal and legal constraints. Sound risk management practices, including asset resiliency, mitigation of risks, and redundancy, will be shared and advanced throughout the sector.
- Cyber Goals and Attributes: Like physical attributes, these assist the ESS in evaluating consequences and vulnerabilities and in developing protec-

tive strategies. Cyber systems that link and help monitor and control the emergency services systems are increasingly recognized as a potential vulnerability. All information that identifies or otherwise describes characteristics of an ESS asset that is created, held, and maintained by the government or the private sector will be protected from unauthorized disclosure according to established procedures, appropriate to the particular level of information.

- Volumetric or Throughput Attributes: These define the extent of any damage, depending on the utilized capacity of the systems, or points where the system may be capacity constrained.
- Personnel Security Goals and Human Attributes: Ensure all personnel directly associated with an emergency services asset are vetted for employment suitability, reliability, and trustworthiness using established processes commensurate with requirements of the respective positions held, in conformance with pertinent security policy. Highly trained and skilled personnel are key factors in a comprehensive ESS security plan. The availability of skilled and experienced technical talent is a concern in the ESS. Sustaining essential technical knowledge is critical to maintaining the sector's safety, reliability, and security.
- Physical Security Goal: Determine the impact or consequence of ESS asset loss, its mission(s) supported, the known or perceived threat, and the vulnerabilities in the asset that the threat is capable of exploiting; identify specific ESS assets the destruction or disruption of which could result in human casualties or economic disruption similar to the effects of weapons of mass destruction; compile a composite of physical security risk assessments for the facility.
- Insider Threat Goal and Attributes: Responsible parties in charge provide security education and training aids to emergency services asset owners/operators not having a security program so that they may implement provisions for the vetting of system and network administrators commensurate with the consequences of the loss of sensitive or classified information, production or provisioning capability, or supply-chain integrity.
- Monitoring and Reporting Goals and Attributes: Ongoing determination of the effectiveness of government threat reporting to officials, owners, and operators responsible for ESS assets, and to local law-enforcement officials and other first responders, including, as appropriate, the medical and mass transportation communities.
- Training and Education Goal and Attributes: Develop and provide continuous specific security education and training materials for ESS asset owner/operators.

HOMELAND SECURITY DIRECTIVES

As a result of 9/11, the Homeland Security Department was formed. On matters pertaining to homeland security, Homeland Security Presidential Directives (HSPDs) are issued by the president. Each directive has a specific meaning and purpose and is carried out by the U.S. Department of Homeland Security. Table 2.1 lists the HSPDs.

Table 2.1. Homeland Security Presidential Directives

HSPD-1: Organization and Operation of the Homeland Security Council. (White House) Ensures coordination of all homeland security–related activities among executive departments and agencies and promotes the effective development and implementation of all homeland security policies.

HSPD-2: Combating Terrorism through Immigration Policies. (White House) Provides for the creation of a task force that will work aggressively to prevent aliens who engage in or support terrorist activity from entering the United States and to detain, prosecute, or deport any such aliens who are within the United States.

HSPD-3: Homeland Security Advisory System. (White House) Establishes a comprehensive and effective means to disseminate information regarding the risk of terrorist acts to federal, state, and local authorities and to the American people.

HSPD-4: National Strategy to Combat Weapons of Mass Destruction. Applies new technologies, increases emphasis on intelligence collection and analysis, strengthens alliance relationships, and establishes new partnerships with former adversaries to counter this threat in all of its dimensions.

HSPD-5: Management of Domestic Incidents. (White House) Enhances the ability of the United States to manage domestic incidents by establishing a single, comprehensive national incident management system.

HSPD-6: Integration and Use of Screening information. (White House) Provides for the establishment of the Terrorist Threat Integration Center.

HSPD-7: Critical Infrastructure Identification, Prioritization, and Protection. (White House) Establishes a national policy for federal departments and agencies to identify and prioritize U.S. critical infrastructure and key resources and to protect them from terrorist attacks.

HSPD-8: National Preparedness. (White House) Identifies steps for improved coordination in response to incidents. This directive describes the way federal departments and agencies will prepare for such a response, including prevention activities during the early stages of a terrorism incident. This directive is a companion to HSPD-5.

HSPD-8 Annex 1: National Planning. Further enhances the preparedness of the United States by formally establishing a standard and comprehensive approach to national planning.

HSPD-9: Defense of United States Agriculture and Food. (White House) Establishes a national policy to defend the agriculture and food system against terrorist attacks, major disasters, and other emergencies.

HSPD-10: Biodefense for the 21st Century. (White House) Provides a comprehensive framework for our nation's biodefense.

HSPD-11: Comprehensive Terrorist-Related Screening Procedures. (White House) Implements a coordinated and comprehensive approach to terrorist-related screening that supports homeland security, at home and abroad. This directive builds upon HSPD-6.

HSPD-12: Policy for a Common Identification Standard for Federal Employees and Contractors. (White House) Establishes a mandatory, government-wide standard for secure and reliable forms of identification issued by the federal government to its employees and contractors (including contractor employees).

HSPD-13: Maritime Security Policy. Establishes policy guidelines to enhance national and homeland security by protecting U.S. maritime interests.

HSPD-14: Domestic Nuclear Detection. Establishes a Domestic Nuclear Detection Office (DNDO) to coordinate efforts to protect the domestic United States against dangers from nuclear or radiological materials. The EPA supports the detection, response, law-enforcement, and information sharing aspects of the DNDO's mission.

HSPD-16: Aviation Strategy. Details a strategic vision for aviation security while recognizing ongoing efforts, and directs the production of a National Strategy for Aviation Security and supporting plans.

HSPD-18: Medical Countermeasures against Weapons of Mass Destruction. (White House) Establishes policy guidelines to draw upon the considerable potential of the scientific community in the public and private sectors to address medical countermeasure requirements relating to CBRN threats.

HSPD-19: Combating Terrorist Use of Explosives in the United States. (White House) Establishes a national policy, and calls for the development of a national strategy and implementation plan, on the prevention and detection of, protection against, and response to terrorist use of explosives in the United States.

HSPD-20: National Continuity Policy. (White House) Establishes a comprehensive national policy on the continuity of federal government structures and operations and a single national continuity coordinator responsible for coordinating the development and implementation of federal continuity policies.

HSPD-21: Public Health and Medical Preparedness. (White House) Establishes a national strategy that will enable a level of public health and medical preparedness sufficient to address a range of possible disasters.

HSPD-23: Cyber Security. Requires federal agencies to monitor cyber activity toward federal agencies' computer systems and, where necessary, provide action to eliminate sources of hostile action. The Environmental Protection Agency (EPA) has a robust security program for both personnel and cyber security as mandated by the directive.

Source: USEPA (2016).

ASSESSING CONSEQUENCES

The potential physical and cyber consequences of any incident, including terror attacks and natural or man-made disasters, are the primary consideration in risk assessment. In the context of this text, consequence is measured as the range of loss or damage that can be expected. The consequences that are considered for national-level comparative risk assessments are based on the criteria set forth in HSPD-7. These criteria can be divided into four main categories:

- Human Impact: Effect on human life and physical well-being (e.g., fatalities, injuries).
- Economic Impact: Direct and indirect effects on the economy (e.g., costs resulting from disruption of products or services, costs to respond to and recover from the disruption, costs to rebuild the asset, and long-term costs due to environmental damage).
- Impact on Public Confidence: Effect on public morale and confidence in national economic and political institutions.
- Impact on Government Capability: Effect on the government's ability to maintain order, deliver minimum essential public services, ensure public health and safety, and carry out national security–related missions.

Moreover, HSPD-7 is important to the ESS in that it required the Department of Homeland Security to "serve as the focal point for the security of cyberspace," with a mission that included "analysis, warning, information sharing, vulnerability reduction, mitigation, and aiding national recovery efforts for critical infrastructure information systems." This directive established a national policy for federal departments and agencies to identify and prioritize U.S. critical infrastructure and key resources and to protect them from terrorist attacks. In addition, it required the heads of all federal agencies to "develop . . . plans for protecting the physical and cyber critical infrastructure and key resources that they own or operate." Hence, the federal government began to directly address issues of cybersecurity within federal government systems (FCC, 2017).

As a result of HSPD-7, the Department of Homeland Security established the National Cybersecurity Division (NCSD). The objectives of this division are "to build and maintain an effective national cyberspace response system, and to implement a cyber-risk management program for protection of critical infrastructure." The primary operational arms of the division are first the Cybersecurity Preparedness and National Cyber Alert System, and secondly the U.S. Computer Emergency Response Team (US-CERT). The National Cyber Alert System was created by US-CERT and DHS to help protect computers. One of US-CERT's overarching goals is to ensure that individuals and agencies have access to timely information through tips and alerts about security topics and events. US-CERT has become the national first line of defense for the war of cybersecurity. CERT's Cyber Risk Management Program assesses risk, prioritizes resources, and executes protective measures in order to secure the cyber infrastructure. It includes such things as current risk assessments and vulnerabilities that are maintained in their vulnerability database, the National Cyber Alert System, for information dissemination, and a number of other references for cybersecurity measures.

In addition to the importance of HSPD-7 providing guidance and direction in cybersecurity and communications protection objectives directly and indirectly related to the critical manufacturing sector (CMS), as a further shot in the security arm, so to speak, HSPD-23 was signed in January 2008 by President Bush; this directive was necessary due to increased cyber activity on an international scale and attacks targeting U.S. computers and networks—including computer-controlled systems. HSPD-7 established a Comprehensive National Cybersecurity Initiative (CNCI). Although the document is classified, public sources have indicated that in addition to establishing the National Cybersecurity Center within the Department of Homeland Security, the initiative had twelve other objectives (FCC, 2017):

- Move toward managing a single federal enterprise network.
- Deploy intrinsic detection systems.
- Develop and deploy intrusion prevention tools.
- Review and potentially redirect research and funding.
- Connect current government cyber operations centers.
- Develop a government-wide cyber intelligence plan.
- Increase the security of classified networks.
- Expand cyber education.
- Define enduring leap-ahead technologies.
- Define enduring deterrent technologies and programs.
- Develop multipronged approaches to supply-chain risk management.
- Define the role of cybersecurity in private-sector domains.

NOTES

Portions of this chapter are adapted from the U.S. Department of Homeland Security Cybersecurity and Infrastructure Security Agency's August 2019 report titled "Emergency Services Sector Landscape" (https://www.cisa.gov/resources-tools/resources/emergency-services-sector-landscape).

1. A. Reichard et al., "Occupational Injuries and Exposures among Emergency Medical Services Workers," *Prehospital Emergency Care* (January 2017).
2. M. D. Weaver et al., "An Observational Study of Shift Length, Crew Familiarity, and Occupational Injury and Illness in Emergency Medical Services Workers," *Occupational and Environmental Medicine* (September 2015).

3. B. Choi et al., "Twenty-Four-Hour Work Shifts, Increased Job Demands, and Elevated Blood Pressure in Professional Firefighters," *International Archives of Occupational and Environmental Health* (July 2016).

4. Federal Emergency Management Agency (FEMA), *U.S. Fire Administration, Firefighter Fatalities in the United States in 2017* (September 2018).

5. SAMHSA, "First Responders: Behavioral Health Concerns, Emergency Response, and Trauma," *Disaster Technical Assistance Center Supplemental Research Bulletin* (May 2018).

6. Centers for Disease Control and Prevention (CDC), National Institute for Occupational Safety and Health (NIOSH), *Emergency Medical Services Workers: Injury Data* (March 2018).

7. H. M. Tiesman et al., "Nonfatal Injuries to Law Enforcement Officers: A Rise in Assaults," *American Journal of Preventive Medicine* (February 2018).

8. CDC, NIOSH, *Emergency Medical Service Workers: Injury Data* (March 2018).

9. *Talos*, "New VPNFilter Malware Targets at Least 500K Networking Devices Worldwide," (May 2018).

10. McAfee Labs, *Threats Report* (April 2017).

11. *The Register*, "World's Biggest DDoS Attack Record Broken after Just Five Days" (March 2018).

12. *Christian Science Monitor*, "911 TDOS near Phoenix, AZ Spread over Twitter" (March 2017).

13. Varonis, *The State of CryptoWall in 2018* (June 2016).

14. Symantec, *Internet Security Threat Report* (April 2018).

15. National Oceanic and Atmospheric Administration (NOAA), "Billion-Dollar Weather and Climate Disasters" (January 2018).

16. U.S. Department of Agriculture (USDA), "Wildland Firefighter Smoke Exposure" (October 2013).

17. Federal Bureau of Investigation (FBI), *Law Enforcement Officers Killed & Assaulted, 2018 (LEOKA)*, Table 24: Law Enforcement Officers Feloniously Killed (Spring 2019).

18. University of Maryland, *Global Terrorism Database*, query for incidents between 2016–2018 in the United States targeting police (April 2019).

19. *Wired*, "Above Devastated Houston, Armies of Drones Prove Their Worth" (September 2017).

20. *New York Times*, "A Closer Look at the Drone Attack on Maduro in Venezuela" (August 2018).

21. American Society of Civil Engineers (ASCE), *2017 Infrastructure Report Card* (March 2017).

22. U.S. Department of Energy (DOE), *Transforming U.S. Energy Infrastructures in a Time of Rapid Change: The First Installment of the Quadrennial Energy Review* (April 2015).

23. U.S. Energy Information Administration (EIA), "Major Utilities Continue to Increase Spending on U.S. Electric Distribution Systems," *Today in Energy* (July

2018); EIA, "Utilities Continue to Increase Spending on Transmission Infrastructure," *Today in Energy* (February 2018).

24. American Water Works Association, *State of the Water Industry Report* (2018).

REFERENCES AND RECOMMENDED READING

Bureau of Labor Statistics. (2014). *Fatal Occupational Injuries, U.S. 2001.*

CDC. (1996). *Violence in the Workplace.*

Census for Fatal Occupational Injury Statistics. (2014). *Injuries, Illnesses, and Fatalities*.https://www.bls.gov/iif/state-data/archive/fatal-occupational-injuries-index-2014.htm, accessed May 17, 2023.

Crayton, J. W. (1983). "Terrorism and the Psychology of the Self." In L. Z. Freedman and Y. Alexander (Eds.), *Perspectives on Terrorism* (pp. 22–41). Wilmington, DE: Scholarly Resources.

DHS. (2003). *The National Strategy for the Physical Protection of Critical Infrastructures and Key Assets.*

DHS. (2015). *Emergency Services Sector Plan*. Washington, DC: Department of Homeland Security. https://www.dhs.gov/critical-manufacturing-sector, accessed May 17, 2023.

DHS. (2017a). *Securing Federal Networks*. https:/www.dhs.gov/topic/securing-federal-networks, accessed May 17, 2023.

DHS. (2017b). *EINSTEIN*. https://www.dhs.gov/einstein, accessed May 17, 2023.

FCC. (2017). "Public Safety Tech Topic #20—Cybersecurity and Communications." https://www.fcc.gov/help/public-safety-tech-topic-20-cyber-security-and-communications, accessed April 14, 2017.

FEMA. (2015). *Protecting Critical Infrastructure against Insider Threats*. https://emilms.fema.gov/is_0915/curriculum/1.html, accessed May 17, 2023.

Ferracuti, F. (1982). "A Sociopsychiatric Interpretation of Terrorism." *Annals of the American Academy of Political and Social Science, 463* (September): 129–41.

FR. (2007). *Federal Register*, 17688–745.

Hudson, R. A. (1999). *The Sociology and Psychology of Terrorism: Who Becomes a Terrorist and Why*? Washington, DC: Library of Congress.

Kelleher, M. D. (1997). *Profiling the Lethal Employee: Case Studies of Violence in the Workplace.* Westport, CT: Greenwood, 1997.

Lees, F. (1996). *Loss Prevention in the Process Industries*, 3:A5.1–A5.11. New York: Butterworth-Heinemann.

Long, D. E. (1990). *The Anatomy of Terrorism*. New York: Free Press.

Margolin, J. (1977). "Psychological Perspectives on Terrorism." In Y. Alexander and S. M. Finger (Eds.), *Terrorism: Interdisciplinary Perspectives*. New York: John Jay Press.

Myers, M., B. Turner, and T. Turner. (1992). *Wayne's World.* Paramount Pictures.

Olson, M. (1971). *The Logic of Collective Action.* Cambridge, MA: Harvard University Press.

OMB. (1998). *Federal Conformity Assessment Activities, Circular A-119*. Washington, DC: White House.

OSHA. (2012). *OSHA Fact Sheet: Workplace Violence*. https://www.ucop.edu/risk-services/_files/bsas/safetymeetings/osha_wkplace_vio_factsheet.pdf, accessed May 17, 2023.

Pearlstein, R. (1991). *The Mind of the Political Terrorist*. Wilmington, DE: Scholarly Resources.

Sullivant, J. (2007). *Strategies for Protecting National Critical Infrastructure Assets: A Focus on Problem-Solving*. New York: Wiley.

USEPA. (2016). *Homeland Security Presidential Directives*.https://www.epa.gov/emergency-response/homeland-security-presidential-directives, accessed May 17, 2023.

Wilkinson, P. (1974). *Political Terrorism*. London: Macmillan.

Chapter 3

Vulnerability Assessment (VA)

One consequence of the events of 9/11 was the Department of Homeland Security's (DHS) directive to establish a Critical Infrastructure Protection Task Force to ensure that activities to protect and secure vital infrastructure are comprehensive and carried out expeditiously. Another consequence is a heightened concern among citizens of the United States over the security of their energy infrastructure (i.e., the uninterrupted supply of electrical power and fuel to power vehicles, homes, manufacturing jobs to pay the rent, etc.) and vital communications systems. As mentioned earlier, along with other critical infrastructure, the emergency services sector is classified as "vulnerable" in the sense that inherent weaknesses in its operating environment could be exploited to cause harm to the system. There is also the possibility of a cascading effect—a chain reaction—due to a terrorist act affecting ESS, which could cause corresponding damage (collateral damage) to other nearby users. In addition to significant damage to the nation's ESS, entities using and needing manufacturing services to function can result in loss of life due to a lack of proper emergency provisions and/or response, and a shutdown of other industries or a loss of equipment necessary for electronic communications and operational control networks could cause catastrophic environmental damage and result in harmful impacts to national security and long-term public health.

ASSESSING VULNERABILITIES

For the purpose of this text and according to FEMA (2008), *vulnerability* is defined as any weakness that can be exploited by an aggressor to make an asset susceptible to hazard damage. In addition, according to the Department of Homeland Security (DHS, 2009), vulnerabilities are physical features or operational attributes that render an entity open to exploitation or susceptible to a given hazard. Vulnerabilities may be associated with physical (e.g., a broken fence), cyber (e.g., lack of a firewall), or human (e.g., untrained guards) factors.

Vulnerability assessments estimate the odds that a characteristic of, or flaw in, an infrastructure could make it susceptible to destruction, disruption, or exploitation based on its design, location, security posture, processes, or operations. Vulnerabilities are typically identified through internal assessments and information sharing with customers, vendors, and suppliers.

A vulnerability assessment methodology was developed as part of the complete Emergency Services Sector-Specific Plan risk assessment methodology. The methodology examined physical, cyber, and human vulnerabilities and considered relevant national preparedness threat scenarios. The process varied depending on the architecture elements being studied and included subject-matter expert interviews, site visits, and modeling and analysis.

The vulnerabilities of ESS elements may vary depending on whether they are operational or implementation specific. Operational vulnerabilities may include those that result from the inherent principles of plant design and location, unanticipated output limitation caused by external factors, or collateral consequences from major disasters or events. Implementation-specific vulnerabilities may be very particular in nature—from deliberately placed bugs in production equipment and protocol deficiencies to back doors in vendor equipment, robotic firmware, or software used in the emergency response process. The magnitude of the implementation vulnerabilities also varies depending on the exposure of the vulnerable equipment. While embedded robotic firmware, for example, may have only limited exposures to configuration and maintenance functions, certain systems require a high degree of exposure in order to provide service to machine operators (DHS, 2009).

Vulnerability assessments are conducted on many levels. DHS has instituted a process to provide awareness training to emergency services asset owner/operators. The purpose of the awareness training is to provide ESS personnel with information about the place of their asset within the overall ESS mission requirements and acquisition process so they will understand their rules and their importance to entities at the corporate and site levels (DHS, 2009). The training focuses on:

- Protection of emergency services interests
- Protection of federal interests
- Importance of facilities fostering relationships with local responders and federal, state, and local law-enforcement/civil authorities for business recovery planning

The awareness training also informs asset owners/operators of the protection measures applied to their proprietary and business-sensitive information provided by and to the ESS. Once ESS assets are identified and prioritized, the next step is to conduct standardized assessments. DHS, working through and with various agencies, has established a standardized mission assurance assessment for application to ESS assets. These assessments consider impact, vulnerability, and threat/hazard (whether from natural disaster, technological failure, human error, criminal activity, or terrorist attack). This approach to risk assessment ensures consideration of relevant factors for each ESS asset and a relative prioritization of risks to support homeland terrorist activities and military operations.

INSIDER THREAT VULNERABILITY

The insider threat to an organization's emergency services activities refers to malicious treatment of equipment and equipment operators and the variety of forms of hostile or intrusive behavior, including worker harassment, workplace violence, etc. The point is that to protect an organization from insider threats, the organization must "use tools to monitor the hiring procedure and monitor the traffic in or out of the computer network. Moreover, the employer must be able to focus inside monitoring on specific people who do something concerning or suspicious; moreover, nontechnical employee behavior must also be monitored" (INSA, 2017).

What monitoring really comes down to, however, is awareness. An important part of any successful vulnerability assessment process is awareness. Security and risk managers (and all employees in general) working in or with the critical infrastructure sectors must be aware of the potential for insider threat vulnerability. Awareness means that personnel within the critical infrastructure sectors must know how to identify and take action against insider threats. To achieve this critical goal, safety and security personnel must be provided with an overview of and be cognizant of common characteristics and indicators associated with malicious insiders and effective measures to counter insider threats.

Protecting against Insider Threats (FEMA, 2015)

As mentioned earlier, when analyzing threats to our nation's critical infrastructure, we tend to focus on malicious actions from boat- or planeloads of outside actors. Of equal concern (and even more so in the author's view) are threats from an insider—someone we have given legitimate access to equipment, information, supporting systems, and resources. The measures we take to detect and protect against external threats may not be sufficient to address threats from insiders.

A malicious insider has access to and inside knowledge of the organization and uses that knowledge with the intent to cause harm. The insider may be a current employee, a former employee, a service provider, or, especially in the current era, a person planted inside who has been brainwashed and radicalized into a terrorist waiting for the moment of maximum impact to people and property.

Given the importance of our nation's critical infrastructure, the actions taken by a malicious employee or service provider could have devastating consequences. Let's look at some actual examples.

- A service provider employee at a nuclear facility stole two five-gallon containers of low-enriched uranium dioxide and then attempted to extort $100,000 by threatening to disperse the material in an unnamed U.S. city.
- A power company field engineer, angry with his supervisor, disabled protection systems at a substation and forced the shutdown of the entire network. More than eight hundred thousand customers lost power as a result.
- Two municipal employees used their access credentials to sabotage the system controlling the traffic lights of a major city, causing widespread traffic delays. The damage took four days to repair.
- A disgruntled supermarket meat-packaging employee intentionally contaminated hamburger meat with a pesticide, causing various levels of illness in ninety-two consumers.

Insider threats endanger the integrity and security of our workplaces and our communities. This section helps you become aware of threat indicators and actions you can take. As stated earlier, and it can't be overstated, awareness is the first step to keeping our nation and workplaces safe.

Insider Threat Defined

The National Infrastructure Advisory Council (2008) defines *insider threat* as follows:

> The insider threat to critical infrastructure is one or more individuals with the access or inside knowledge of a company, organization, or enterprise that would allow them to exploit the vulnerabilities of the entity's security, systems, services, products, or facilities with the intent to cause harm.

A person who takes advantage of access or inside knowledge in such a manner is commonly referred to as a "malicious insider."

The Scope of Insider Threats

Insider threats can be accomplished through either physical or cyber means and may involve any of the following:

- Physical or Information Technology Sabotage: Involves modification or damage to an organization's facilities, property, assets, inventory, or systems with the purpose of harming or threatening harm to an individual, the organization, or the organization's operations.
- Theft of Intellectual Property: Involves removal or transfer of an organization's intellectual property outside the organization through physical or electronic means (also known as economic espionage).
- Theft of Economic Fraud: Involves acquisition of an organization's financial or other assets through theft or fraud.
- National Security Espionage: Involves obtaining information or assets with a potential impact on national security through clandestine activities.

Common Characteristics and Traits of Malicious Insiders

Based on research conducted by the Software Engineering Institute at Carnegie Mellon University and the U.S. Secret Service National Threat Assessment Center, malicious insiders often are perceived or known to be difficult or high-maintenance employees who are:

- Obviously unhappy or extremely resentful
- Having financial, performance, or behavioral problems
- At risk (or perceived to be) for layoff or termination

Keep in mind that not all malicious insiders fit this characterization. Insiders involved in national security espionage, for example, may exhibit few outward signs. In the majority of cases, however, management and/or human resources personnel were well aware of the employees and their issues prior to an incident.

Personal Factors Associated with Insiders

The following motives and personal situations are frequently linked with malicious insiders:

Personal or Behavioral Problems

- Vulnerable to blackmail
- Experiencing family or financial problems
- Prone to compulsive or destructive behavior
- Subject to ego or self-image issues

Personal Desires

- Seeking adventure or thrill
- Seeking approval and returned favors
- Professing allegiance

Workplace Issues

- Experiencing problems at work
- Feeling anger or a need for revenge

Organizational Factors That Embolden Malicious Insiders

The following organizational factors have been known to encourage or present opportunities to potential malicious insiders.

Access and Availability

- Ease of access to materials and information
- Ability to exit the facility or network with materials or information

Policies and Procedures

- Undefined or inadequate policies and procedures
- Inadequate training
- Lack of training

Time Pressure and Consequences

- Rushed employees
- Perception of lack of consequences

Insider Activities and Behavior You May See

Insider threats may be detected through particular activities and behavior on the part of the insider. This section of the presentation identifies those indicators. These activities and behaviors often will appear unusual or suspicious. Keep in mind that there may be several explanations for a particular activity or behavior identified here, but when combined with other factors, certain activities or behavior point toward a possible insider threat. A combination or confluence of indicators should not be ignored.

Types of Insider Activities and Behavior

Unusual or suspicious insider activities and behavior can be described using the following categories:

- Inappropriate interest or acquisition
- Unauthorized or unusual computer use
- Unusual hours, contacts, or travel
- Secretive or peculiar behavior
- Personal or financial issues

Employer Actions

Employer action is required to ensure and maintain security concerns.

- Clearly communicating and consistently enforcing security policies and controls
- Ensuring that proprietary information and materials are adequately, if not robustly, protected
- Routinely monitoring computer networks for suspicious activity
- Ensuring that security (to include computer network security) personnel have the tools they need
- Consulting with legal and law-enforcement experts as needed to ensure compliance with the law

Employee Actions

Critical infrastructure organizations today employ a number of security measures to reduce the risk of insider threats. The measures involving employees include, but are not limited to:

- Using appropriate screening processes to select new employees
- Educating employees about security or other protocols
- Encouraging and providing nonthreatening, convenient ways for employees to report suspicious activity in a confidential manner
- Becoming familiar with behavior and activities associated with malicious insiders
- Documenting and evaluating incidents of suspicious or disruptive behavior
- Consulting with legal and law-enforcement experts as needed to ensure compliance with the law

THE VULNERABILITY ASSESSMENT (VA)

A *vulnerability assessment*[1] involves an in-depth analysis of the facility's functions, systems, and site characteristics to identify facility weaknesses and lack of redundancy, and to determine mitigations or corrective actions that can be designed and implemented to reduce those vulnerabilities. A vulnerability assessment can be a stand-alone process or part of a full risk assessment. During this assessment, the analysis of site assets is based on: (a) the identified threat, (b) the criticality of the assets, and (c) the level of protection chosen (i.e., based on willingness or unwillingness to accept risk).

It is important to point out that post-9/11, all sectors have taken great strides to protect their critical infrastructure. For instance, government and industry have developed vulnerability assessment methodologies for critical infrastructure systems and trained thousands of auditors and others to conduct them.

The actual complexity of vulnerability assessments will range based upon the design and operation of the emergency services asset. The nature and extent of the VA will differ among systems based on a number of factors, including system size, potential population at risk, knowledge about types of threats, available security technologies, and applicable local, state, and federal regulations. Preferably, a VA will be "performance based," meaning that it evaluates the risk to emergency services assets based on the effectiveness (performance) of existing and planned measures to counteract adversarial actions. According to USEPA (2002), the common elements of CMS vulnerability assessments are as follows:

- Characterization of the emergency services sector, including its mission and objectives
- Identification and prioritization of adverse consequences to avoid

- Determination of critical assets that might be subject to malevolent acts that could result in undesired consequences
- Assessment of the likelihood (qualitative probability) of such malevolent acts from adversaries
- Evaluation of existing countermeasures
- Analysis of current risk and development of a prioritized plan for risk reduction

Benefits of Assessments

ESS members should routinely perform vulnerability assessments to better understand threats and vulnerabilities, determine acceptable levels of risk, and stimulate action to mitigate identified vulnerabilities. The direct benefits of performing a vulnerability assessment include:

- Building and Broadening Awareness: The assessment process directs senior management's attention to security. Security issues, risks, vulnerabilities, mitigation options, and best practices are brought to the surface. Awareness is one of the least expensive and most effective methods for improving the organization's overall security posture.
- Establishing or Evaluating against a Baseline: If a baseline has been previously established, an assessment is an opportunity for a checkup to gauge the improvement or deterioration of an organization's security posture. If no previous baseline has been established (or the work was not uniform or comprehensive), an assessment is an opportunity to integrate and unify previous efforts, define common metrics, and establish a definitive baseline. The baseline can also be compared against best practices to provide perspective on an organization's security posture.
- Identifying Vulnerabilities and Developing Responses: Generating lists of vulnerabilities and potential responses is usually a core activity and outcome of an assessment. Sometimes, due to budget, time, complexity, and risk considerations, the response selected for many of the vulnerabilities may be nonaction, but after completing the assessment process, these decisions will be conscious ones, with a documented decision process and item-by-item rationale available for revisiting issues at scheduled intervals. This information can help drive or motivate the development of a risk management process.
- Categorizing Key Assets and Driving the Risk Management Process: An assessment can be a vehicle for reaching corporate-wide consensus on a hierarchy of key assets. This ranking, combined with threat, vulnerability,

and risk analysis, is at the heart of any risk management process. For many organizations, the Y2K bug was the first time a company-wide inventory and ranking of key assets was attempted. An assessment allows any organization to revisit that list from a broader and more comprehensive perspective.

- Developing and Building Internal Skills and Expertise: A security assessment, when not implemented in an "audit" mode, can serve as an excellent opportunity to build security skills and expertise within an organization. A well-structured assessment can have elements that serve as a forum for crosscutting groups to come together and share issues, experiences, and expertise. External assessors can be instructed to emphasize "teaching and collaborating" rather than "evaluating" (the traditional role). Whatever the organization's current level of sophistication, a long-term goal should be to move the organization toward a capability for self-assessment.
- Promoting Action: Although disparate security efforts may be underway in an organization, an assessment can crystallize and focus management attention and resources on solving specific and systemic security problems. Often the people in the trenches are well aware of security issues (and even potential solutions) but are unable to convert their awareness to action. An assessment provides an outlet for their concerns and the potential to surface these issues at appropriate levels (legal, financial, executive) and achieve action. An assessment that is well designed and executed does not only identify vulnerabilities and make recommendations; it also gains executive buy-in, identifies key players, and establishes a set of crosscutting groups that can convert those recommendations into action.
- Kicking Off an Ongoing Security Effort: An assessment can be used as a catalyst to involve people throughout the organization in security issues, build crosscutting teams, establish permanent forums and councils, and harness the momentum generated by the assessment to build an ongoing institutional security effort. The assessment can lead to the creation of either an actual or a virtual (matrixed) security organization.

Vulnerability Assessment Process

Table 3.1 provides an overview of the elements included in the assessment methodology. The elements included in this overview are based on actual in-field experience and lessons learned.

Table 3.1. Basic Elements in Vulnerability Assessments

Element	*Points to Consider*
1. Characterization of the communications entity, including its mission and objectives.	What are the important missions of the system to be assessed? Define the highest-priority services provided by the utility. Identify the industry's customers: • General public • Government • Military • Industrial • Critical care • Retail operations • Firefighting What are the most important facilities, processes, and assets of the system for achieving the mission objectives and avoiding undesired consequences? Describe the: • Industry facilities • Operating procedures • Management practices that are necessary to achieve the mission objectives • How the industry operates • Treatment processes • Storage methods and capacity • Energy use and storage • Distribution system In assessing those assets that are critical, consider critical customers, dependence on other infrastructures (e.g., chemical, transportation, communications), contractual obligations, single points of failure, chemical hazards and other aspects of the industry's operations, or availability of industry utilities that may increase or decrease the criticality of specific facilities, processes, and assets.
2. Identification and prioritization of adverse consequences to avoid.	Take into account the impacts that could substantially disrupt the ability of the system to provide a safe and reliable supply of materials. Emergency services sector systems should use the vulnerability assessment process to determine how to reduce risk associated with the consequences of significant concern. Ranges of consequences or impacts for each of these events should be identified and defined. Factors to be considered in assessing the consequences may include: • Magnitude of service disruption • Economic impact (such as replacement and installation costs for damaged critical assets or loss of revenue due to service outage) • Number of illnesses or deaths resulting from an event • Impact on public confidence in the material supply

(*continued*)

Table 3.1. *Continued*

Element	*Points to Consider*
	• Chronic problems arising from specific events • Other indicators of the impact of each event as determined by the emergency services sector Risk reduction recommendations at the conclusion of the vulnerability assessment strive to prevent or reduce each of these consequences.
3. Determination of critical assets that might be subject to malevolent acts that could result in undesired consequences.	What are malevolent acts that could reasonably cause undesired consequences? • Electronic, computer, or other automated systems that are utilized by emergency services sector entities (e.g., supervisory control and data acquisition [SCADA]) • The use, storage, or handling of various manufacturing supplies • The operation and maintenance of such systems
4. Assessment of the likelihood (qualitative probability) of such malevolent acts by adversaries.	Determine the possible modes of attack that might result in consequences of significant concern based on critical assets of the emergency services sector entity. The objective of this step of the assessment is to move beyond what is merely possible and determine the likelihood of a particular attack scenario. This is a very difficult task as there is often insufficient information to determine the likelihood of a particular event with any degree of certainty. The threats (the kind of adversary and the mode of attack) selected for consideration during a vulnerability assessment will dictate, to a great extent, the risk reduction measures that should be designed to counter the threat(s). Some vulnerability assessment methodologies refer to this as a "design-basis threat" (DBT) where the threat serves as the basis for the design of countermeasures, as well as the benchmark against which vulnerabilities are assessed. It should be noted that there is no single DBT or threat profile for all emergency services systems in the United States. Differences in geographic location, size of the utility, previous attacks in the local area, and many other factors will influence the threat(s) that the emergency services sector entity should consider in their assessments. Emergency services sector entities should consult with local FBI and/or other law-enforcement agencies, public officials, and others to determine the threats upon which their risk reduction measures should be based.

Element	*Points to Consider*
5. Evaluation of existing countermeasures. (Depending on countermeasures already in place, some critical assets may already be sufficiently protected. This step will aid in identification of the areas of greatest concern and help to focus priorities for risk reduction.)	What capabilities does the system currently employ for detection, delay, and response? • Identify and evaluate current detection capabilities such as intrusion detection systems, energy quality monitoring, operational alarms, guard post orders, and employee security awareness programs. • Identify current delay mechanisms such as locks and key control, fencing, structural integrity of critical assets, and vehicle access checkpoints. • Identify existing policies and procedures for evaluation and response to intrusion and system malfunction alarms and to cyber system intrusions. It is important to determine the performance characteristics. Poorly operated and maintained security technologies provide little or no protection. What cyber-protection system features does the facility have in place? Assess what protective measures are in place for the SCADA and business-related computer information systems such as: • Firewalls • Modem access • Internet and other external connections, including wireless data and voice communications • Security policies and protocols It is important to identify whether vendors have access rights and/or "back doors" to conduct system diagnostics remotely. What security policies and procedures exist, and what is the compliance record for them? Identify existing policies and procedures concerning: • Personal security • Physical security • Key and access badge control • Control of system configuration and operational data • Vendor deliveries • Security training and exercise records

(*continued*)

Table 3.1. ***Continued***

Element	*Points to Consider*
6. Analysis of current risk and development of a prioritized plan for risk reduction.	Information gathered on threat, critical assets, emergency services sector operations, consequences, and existing countermeasures should be analyzed to determine the current level of risk. The utility should then determine whether current risks are acceptable or risk reduction measures should be pursued. Recommended actions should measurably reduce risks by reducing vulnerabilities and/or consequences through improved deterrence, delay, detection, and/or response capabilities or by improving operational policies or procedures. Selection of specific risk reduction actions should be completed prior to considering the cost of the recommended action(s). Facilities should carefully consider both short- and long-term solutions. An analysis of the cost of short- and long-term risk reduction actions may impact which actions the utility chooses to achieve its security goals. Facilities may also want to consider security improvements. Security and general infrastructure may provide significant multiple benefits. For example, improved treatment processes or system redundancies can both reduce vulnerabilities and enhance day-to-day operation. Generally, strategies for reducing vulnerabilities fall into three broad categories: • Sound business practices—affect policies, procedures, and training to improve the overall security-related culture at the chemical facility. For example, it is important to ensure that rapid communication capabilities exist between public health authorities and local law enforcement and emergency responders. • System upgrades—include changes in operations, equipment, processes, or infrastructure itself that make the system fundamentally safer. • Security upgrades—improve capabilities for detection, delay, or response.

In table 3.1, step 3 deals with the identification of asset criticality. This is an important step in any vulnerability assessment. Identifying asset criticality serves several functions:

- It enables more careful consideration of factors that affect risk, including threats, vulnerabilities, and consequences of loss or compromise of the asset.
- It enables more focused and thorough consideration of loss or compromise of the asset.
- It enables leaders to develop robust methods for managing consequences of asset loss (restoration).
- It provides a means to increase awareness of a broad range of employees to protect truly critical assets and to differentiate in policies and procedures the heightened protection they require.

As previously indicated, identifying the criticality of assets is used primarily to focus the efforts of a vulnerability analysis. It also assists in the ranking of various recommendations for reducing vulnerabilities. As an example, let's take a look at the criticality of electric power assets and operations included in the normal operation of ESS members:

Physical

- Generators
- Substations
- Transformers
- Transmission lines
- Distribution lines
- Control center
- Warehouses
- Office buildings
- Internal and external infrastructure dependencies
- Manufacturing and process equipment

Cyber

- SCADA systems
- Networks
- Databases
- Business systems
- Telecommunications

Interdependencies

- Single-point nodes of failures
- Critical infrastructure components of high reliance

VULNERABILITY ASSESSMENT METHODOLOGY

Vulnerability assessment methodology consists of ten elements. Each element, along with a description of each, is listed below (USDOE, 2002).

1. Network architecture
2. Threat environment
3. Penetration testing
4. Physical security
5. Physical asset analysis
6. Operations security
7. Policies and procedures
8. Impact analysis
9. Infrastructure interdependencies
10. Risk characterization

Network Architecture

This element provides an analysis of the information assurance features of the information network(s) associated with the organization's critical information systems. Information examined should include network topology and connectivity (including subnets), principal information assets, interface and communication protocols, the function and lineage of major software and hardware components (especially those associated with information security such as intrusion detectors), and policies and procedures that govern security features of the network.

Procedures for information assurance in the system, including authentication of access and management of access authorization, should be reviewed. The assessment should identify any obvious concerns related to architectural vulnerabilities as well as operating procedures. Existing security plans should be evaluated, and the results of any prior testing should be analyzed. Results from the network architecture assessment should include potential recommendations for changes in the information architecture, functional areas and categories where testing is needed, and suggestions regarding system design that would enable more effective information and information system protection.

Three techniques are often used in conducting the network architecture assessment:

1. Analysis of network and system documentation during and after the site visit
2. Interview with facility staff, managers, and the chief information officer
3. Tours and physical inspections of key facilities

Threat Environment

Development of a clear understanding of the threat environment is a fundamental element of risk management. When combined with an appreciation of the value of the information assets and systems, and with awareness of the impact of unauthorized access and subsequent malicious activity, an understanding of threats provides a basis for better defining the level of investment needed to prevent such access.

The threat of a terrorist attack to ESS infrastructure is real and could come from several sources, including physical, cyber, and interdependency. In addition, threats could come from individuals or organizations motivated by financial gain or persons who derive pleasure from such penetration (e.g., recreational hackers, disgruntled employees). Other possible sources of threats are those who want to accomplish extremist goals (e.g., environmental terrorists, antinuclear advocates) or to embarrass one or more organizations.

This element should include a characterization of these and other threats, identification of trends in these threats, and ways in which vulnerabilities are exploited. To the extent possible, characterization of the threat environment should be localized; that is, it should be located within the organization's service area.

Penetration Testing

The purpose of network penetration testing is to utilize active scanning and penetration tools to identify vulnerabilities that a determined adversary could easily exploit. Penetration testing can be customized to meet the specific needs and concerns of the ESS unit. In general, penetration testing should include a test plan and details on the rules of engagement (ROE). It should also include a general characterization of the access points of the critical information systems and communication interface connections, such as modem network connections, access points to principal network routers, and other external connections important to emergency services. Finally, penetration testing should include identified vulnerabilities and, in particular, whether access could be gained to the control network or to specific subsystems or devices that have a critical role in ensuring continuity of service.

Penetration testing consists of an overall process of establishing the ground rules or ROE for the test; establishing a white cell for continuous communication; developing a format or methodology for the test; conducting the test; and generating a final report that details methods, findings, and recommendations.

Penetration testing methodology consists of three phases: reconnaissance, scenario development, and exploitation. A one-time penetration test can provide the utility with valuable feedback; however, it is far more effective if performed on a regular basis. Repeated testing is recommended because new threats develop continuously, and the networks, computers, and architecture of the communications sector unit or utility are likely to change over time.

Physical Security

A critical dependency for the ESS as well as for other sectors is related to the physical security of facilities. The purpose of physical security assessment is to examine and evaluate the systems in place (or being planned) and to identify potential improvements in this area for the sites evaluated. Physical security systems include access controls, barriers, locks and keys, badges and passes, intrusion detection devices and associated alarm reporting and displays, closed-circuit television (assessment and surveillance), communications equipment (telephone, two-way radio, intercom, cellular), lighting (interior and exterior), power sources (line, battery, generator), inventory control, postings (signs), security system wiring, and protective force. Physical security systems are reviewed for design, installation operation, maintenance, and testing.

The physical security assessment should focus on those sites directly related to the critical facilities, including information systems and assets required for operation. Typically included are facilities that house critical equipment or information assets or networks dedicated to the operation of electric, oil, or gas transmission, storage, or delivery systems. Other facilities can be included on the basis of criteria specified by the organization being assessed. Appropriate levels of physical security are contingent upon the value of company assets, the potential threats to those assets, and the cost associated with protecting the assets. Once the cost of implementing/maintaining physical security programs is known, it can be compared to the value of the company assets, thus providing the necessary information for risk management decisions. The focus of the physical security assessment task is determined by prioritizing the company assets; that is, the most critical assets receive the majority of the assessment activity.

At the start of the assessment, survey personnel should develop a prioritized listing of company assets. This list should be discussed with company personnel to identify areas of security strengths and weaknesses. During these initial interviews, assessment areas that would provide the most benefit to the company should be identified; once known, they should become the major focus of the assessment activities.

The physical security assessment of each focus area usually consists of the following:

- Physical security program (general)
- Physical security program (planning)
- Barriers
- Access controls/badges
- Locks/keys
- Intrusion detection systems
- Communications equipment
- Protective force/local law-enforcement agency

The key to reviewing the above topics is not to just identify whether they exist but to determine the appropriate level that is necessary and consistent with the value of the asset being protected. Physical security assessment worksheets provide guidance on appropriate levels of protection.

Once the focus and content of the assessment task have been identified, the approach to conducting the assessment can be either at the "implementation level" or at the "organizational level." The approach taken depends on the maturity of the security program.

For example, a company with a solid security infrastructure (staffing plans/procedures, funding) should receive a cursory review of these items; however, facilities where the security programs are being implemented should receive a detailed review. The security staff can act upon deficiencies found at the facilities once reported.

For companies with an insufficient security organization, the majority of the time spent on the assessment should take place at the organizational level to identify the appropriate staffing/funding necessary to implement security programs to protect company assets. Research into specific facility deficiencies should be limited to finding just enough examples to support any staffing/funding recommendations.

Physical Asset Analysis

The purpose of the physical asset analysis is to examine the systems and physical operational assets to ascertain whether vulnerabilities exist. Included in this element is an examination of asset utilization, system redundancies, and emergency operating procedures. Consideration should also be given to the topology and operating practices for electric and gas transmission, processing, storage, and delivery, looking specifically for those elements that either singly or in concert with other factors provide a high potential for disrupting service. This portion of the assessment determines company and industry trends regarding these physical assets. Historic trends, such as asset utilization, maintenance, new infrastructure investments, spare parts, SCADA linkages, and field personnel, are part of the scoping element.

The proposed methodology for physical assets is based on a macro-level approach. The analysis can be performed with company data, public data, or both. Some companies might not have readily available data or might be reluctant to share that data.

Key output from the analysis should be graphs that show trends. The historic data analysis should be supplemented with on-site interviews and visits. Items to focus on during a site visit include the following:

- Trends in field testing
- Trends in maintenance expenditures
- Trends in infrastructure investments
- Historic infrastructure outages
- Critical system components and potential system bottlenecks
- Overall system operation controls
- Use and dependency of SCADA systems
- Linkages of operation staff with physical and IT security
- Adequate policies and procedures
- Communications with other regional utilities
- Communications with external infrastructure providers
- Adequate organizational structure

Operations Security

Operations security (OPSEC) is the systematic process of denying potential adversaries (including competitors or their agents) information about the capabilities and intentions of the host organization. OPSEC involves identifying, controlling, and protecting generally nonsensitive activities concerning planning and execution of sensitive activities. The OPSEC assessment reviews the processes and practices employed for denying adversary access to sensitive and nonsensitive information that might inappropriately aid or abet an individual's or organization's disproportionate influence over system operation. This assessment should include a review of security training and awareness programs, discussions with key staff, and tours of appropriate principal facilities. Information that might be available through public access should also be reviewed.

Policies and Procedures

The policies and procedures by which security is administered (1) provide the basis for identifying and resolving issues; (2) establish the standards of reference for policy implementation; and (3) define and communicate roles, responsibilities, authorities, and accountabilities for all individuals and organizations that interface with critical systems. They are the backbone for decisions and day-to-day security operations. Security policies and procedures become particularly important at times when multiple parties must interact to effect a desired level of security and when substantial legal ramifications could result from policy violations. Policies and procedures should be reviewed to determine whether they (1) address the key factors affecting security; (2) enable effective compliance, implementation, and enforcement; (3) reference or conform to established standards; (4) provide clear and comprehensive guidance; and (5) effectively address roles, responsibilities, accountabilities, and authorities.

The objective of the policies and procedures assessment task is to develop a comprehensive understanding of how a facility protects its critical assets through the development and implementation of policies and procedures. Understanding and assessing this area provides a means of identifying strengths and areas for improvement that can be achieved through:

- Modification of current policies and procedures
- Implementation of current policies and procedures
- Development and implementation of new policies and procedures
- Assurance of compliance with policies and procedures
- Cancellation of policies and procedures that are no longer relevant, or are inappropriate, for the facility's current strategy and operations

Impact Analysis

A detailed analysis should be conducted to determine the influence that exploitation of unauthorized access to critical facilities or information systems might have on an organization's operations (e.g., market and/or physical operations). In general, such an analysis would require thorough understanding of (1) the applications and their information processing, (2) decisions influenced by this information, (3) independent checks and balances that might exist regarding information upon which decisions are made, (4) factors that might mitigate the impact of unauthorized access, and (5) secondary impacts of such access.

Similarly, the physical chain of events following disruption, including the primary, secondary, and tertiary impacts of disruption, should be examined.

The purpose of the impact analysis is to help estimate the impact that detrimental impacts could have on ESS units. The impact analysis provides an introduction to risk characterization by providing quantitative estimates of these impacts so that the ESS unit can implement a risk management program and weigh the risks and costs of various mitigation measures.

Infrastructure Interdependencies

The term *infrastructure interdependencies* refers to the physical and electronic (cyber) linkages within emergency services and among our nation's critical infrastructures—energy (electric power, oil, natural gas), communications, telecommunications, transportation, water supply systems, banking and finance, manufacturing, and government services. This task identifies the direct infrastructure linkages between and among the infrastructures that support critical facilities as recognized by the organization. Performance of this task requires a detailed understanding of an organization's functions, internal infrastructures, and how these link to external infrastructures.

The purpose of the infrastructure interdependencies assessment is to examine and evaluate the infrastructures (internal and external) that support critical facility functions, along with their associated interdependencies and vulnerabilities.

Risk Characterization

Risk characterization provides a framework for prioritizing recommendations across all task areas. The recommendations for each task area are judged against a set of criteria to help prioritize the recommendations and assist the organization in determining the appropriate course of action. It provides a framework for assessing vulnerabilities, threats, and potential impacts (determined in the other tasks). In addition, the existing risk analysis and management process at the organization should be reviewed and, if appropriate, utilized for prioritizing recommendations. The degree to which corporate risk management includes security factors is also evaluated.

VULNERABILITY ASSESSMENT PROCEDURES

Vulnerability assessment procedures can be conducted for ESS assets using various methodologies. For example, the checklist analysis is an effective technology. In addition, Pareto analysis (80/20 principle), relative ranking, pre-removal risk assessment (PRRA), change analysis, failure mode and effects analysis (FMEA), fault-tree analysis, event-tree analysis, what-if analysis, and hazard and operability (HAZOP) analysis can be used in conducting the assessment.

Based on personal experience, the what-if analysis and HAZOP analysis seem to be the most user-friendly methodologies to use. A sample what-if analysis procedural outline is presented below, followed by a brief explanation and outline for conducting HAZOP.

What-If Analysis Procedure/Sample What-If Questions

The steps in a what-if checklist analysis are as follows:

1. Select the team (personnel experienced in the process).
2. Assemble information (piping and instrumentation diagrams [P&IDs], process flow diagrams [PFDs], operating procedures, equipment drawings, etc.).
3. Develop a list of what-if questions.
4. Assemble your team in a room where each team member can view the information.
5. Ask each what-if question in turn and determine:
    - What can cause the deviation from design intent that is expressed by the question?
    - What adverse consequences might follow?
    - What are the existing design and procedural safeguards?
    - Are these safeguards adequate?
    - If these safeguards are not adequate, what additional safeguards does the team recommend?
6. As the discussion proceeds, record the answers to these questions in tabular format.
7. Do not restrict yourself to the list of questions that you developed before the project started. The team is free to ask additional questions at any time.
8. When you have finished the what-if questions, proceed to examine the checklist. The purpose of this checklist is to ensure that the team has not forgotten anything. While you are reviewing the checklist, other what-if questions may occur to you.
9. Make sure that you follow up all recommendations and action items that arise from the hazards evaluation.

HAZOP Analysis

The HAZOP analysis technique uses a systematic process to (1) identify possible deviations from normal operations and (2) ensure that safeguards are in place to help prevent accidents. HAZOP analysis uses special adjectives (such as *speed*, *flow*, *pressure*, etc.) combined with process conditions (such as *more*, *less*, *no*, etc.) to systematically consider all credible deviations from normal conditions. The adjectives, called guide words, are a unique feature of HAZOP analysis.

In this approach, each guide word is combined with relevant process parameters and applied at each point (study node, process section, or operating step) in the process that is being examined:

Guide Word	Meaning
No	Negation of the design intent
Less	Quantitative decrease
More	Quantitative increase
Part of	Other materials present by intent
As well as	Other materials present unintentionally
Reverse	Logical opposite of the intent
Other than	Complete substitution

Common HAZOP Analysis Process Parameters:

Flow	Time	Frequency	Mixing
Pressure	Composition	Viscosity	Addition
Temperature	pH	Voltage	Separation
Level	Speed	Information	Reaction

The following is an example of creating deviations using guide words and process parameters:

Guide Word		Parameter		Deviation
NO	+	FLOW	=	NO FLOW
MORE	+	PRESSURE	=	HIGH PRESSURE
AS WELL AS	+	ONE PHASE	=	TWO PHASE
OTHER THAN	+	OPERATION	=	MAINTENANCE
MORE	+	LEVEL	=	HIGH LEVEL

Guide words are applied to both the more general parameters (e.g., react, mix) and the more specific parameters (e.g., pressure, temperature). With the general parameters, it is not unusual to have more than one deviation from the application of one guide word. For example, "more reaction" could mean either that a reaction takes place at a faster rate or that a greater quantity of product results. On the other hand, some combinations of guide words and parameters will yield no sensible deviation (e.g., "as well as" with "pressure").

HAZOP Procedure

1. Select the team.
2. Assemble information (P&IDs, PFDs, operating procedures, equipment drawings, etc.).
3. Assemble your team in a room where each team member can view P&IDs.
4. Divide the system you are reviewing into nodes (you can present the nodes, or the team can choose them as you go along).
5. Apply appropriate deviations to each node. For each deviation, address the following questions:
    - What can cause the deviation from design intent?
    - What adverse consequences might follow?
    - What are the existing design and procedural safeguards?
    - Are these safeguards adequate?
    - If these safeguards are not adequate, what does the team recommend?
6. As the discussion proceeds, record the answers to these questions in tabular format.

VULNERABILITY ASSESSMENT: CHECKLIST PROCEDURE

In performing the vulnerability assessment of any critical ESS unit or facility, one of the simplest methodologies to employ is the checklist. The "Building Vulnerability Assessment Checklist," developed by the Department of Veterans Affairs (VA) and part of FEMA 426, *Reference Manual to Mitigate Potential Terrorist Attacks against Buildings*, is the reference manual that is recommended in this book.[2] It is an excellent guide for conducting a viable checklist-type vulnerability assessment. This checklist will help you to prepare your threat assessment because it allows a consistent security evaluation of designs at various levels. The checklist can be used as a screening tool for preliminary design vulnerability assessment, and it supports all steps for use by assessment teams during preparation for interviews with facility representatives to ensure that all relevant aspects of the ESS assets are considered in the survey.

NOTES

1. Much of the information in this section is from the U.S. Department of Energy, *Vulnerability Assessment Methodology: Electric Power Infrastructure* (Washington, DC, 2002); U.S. Army Research Laboratory, *Vulnerability Risk Assessment, ARL-TR-1045* (Washington, DC: Department of Defense, 2000); U.S. Department of Justice, *A Method to Assess the Vulnerability of U.S. Chemical Facilities* (Washington, DC, 2002).

2. This manual is available on the internet at https://www.dhs.gov/science-and-technology/bips-06fema-426-reference-manual-mitigate-potential-terrorist-attacks-against, accessed May 17, 2023.

REFERENCES AND RECOMMENDED READING

CBO. (2004). *Homeland Security and the Private Sector*.https://www.cbo.gov/sites/default/files/cbofiles/ftpdocs/60xx/doc6042/12-20-homelandsecurity.pdf, accessed May 17, 2023.

CSI. (2011). *2010/2011 Computer Crime and Security Survey*. Orlando, FL: Computer Security Institute.

DHS. (2009). *National Infrastructure Protection Plan*. https://www.dhs.gov/xlibrary/assets/NIPP_Plan.pdf, accessed May 17, 2023.

DHS. (2017). *Emergency Services Sector*. Washington, DC: Department of Homeland Security. https://www.dhs.gov/emergency-services-sector, accessed May 17, 2023.

DOD. (2010). *DoD Policy and Responsibility for Critical Infractions—DODD 3020.40*. Washington, DC: U.S. Department of Defense.

FBI. (2012). *Insider Espionage. Report before the House Committee on Homeland Security Subcommittee on Counter Terrorism and Intelligence*. Washington, DC: Federal Bureau of Investigation.

FEMA. (2008). *FEMA 452: Risk Assessment: A How-To Guide*. https://www.wbdg.org/FFC/DHS/ARCHIVES/fema452.pdf, accessed May 17, 2023.

FEMA. 2015. *Protecting Critical Infrastructure against Insider Threats.* https://emilms.fema.gov/is_0915/curriculum/1.html, accessed May 17, 2023.

INSA. (2017). *Building a Stronger Intelligence Community*. Arlington, VA: Intelligence and National Security Alliance. https://www.insaonline.org/For-Sorting/Content-Since-2022-04-01-Ran-On-2022-04-01/location-not-determined/membership-534, accessed May 17, 2023.

National Infrastructure Advisory Council. (2008). *First Report and Recommendations on the Insider Threat to Critical Infrastructure.* Washington, DC.

Spellman, F. R. (1997). *A Guide to Compliance for Process Safety Management/Risk Management Planning (PSM/RMP)*. Lancaster, PA: Technomic Publishing.

USDOE. (2010). *Energy Sector-Specific Plan: An Annex to the National Infrastructure Protection Plan.* Washington, DC: U.S. Department of Energy.

USDOE. (2002). *Vulnerability Assessment Methodology: Electric Power Infrastructure*. Washington, DC: U.S. Department of Energy.

USEPA. (2002). *Vulnerability Assessment Fact Sheet*. EPA 816-F-02-025. https://nepis.epa.gov/Exe/ZyPDF.cgi/P1004AYS.PDF?Dockey=P1004AYS.PDF, accessed May 17, 2023.

Chapter 4

Preparation

Tools and Responses

Because of the seriousness of the threat of terrorism to the nation's communications sector and the enormous economic and security implications of such attacks, USFCC, USDOE, USEPA and other agencies have worked nonstop since 9/11 to gather and provide as much advice and guidance as possible to aid ESS personnel in protecting ESS assets and associated critical support infrastructure. In this chapter, we provide an overview of important tools that can be used to assist the ESS in guarding against the threat of terrorism. In the discussion, keep in mind that even though we may be addressing a manufacturing sector issue the guidance provided therein could be used to protect the other critical infrastructure sectors.

THREATS AND INCIDENTS

Based on evidence of losses from past accidents, indications of the potential for human and environmental losses and economic costs from an attack on an ESS facility or producer comes from major incidents that have occurred both abroad and in the United States. Those events indicate that the human and environmental losses could be significant (CBO, 2004).

ESS threats and incidents may be of particular concern due to the range of potential consequences:

- Creating an adverse impact on public health within a population
- Disrupting system operations and interrupting the supply of critical military components
- Causing physical damage to system infrastructure
- Reducing public confidence in the manufacturing supply system
- Long-term denial of basic security and protection and the cost of replacement

Keep in mind that some of these consequences would only be realized in the event of a successful terrorist incident; however, the mere threat of terrorism can also have an adverse impact on industries that depend on a safe, steady supply of manufactured goods and associated supplies. In addition, the economic implications of such attacks are potentially enormous. For example, many believe that the reason we are looking at oil at more than $60 a barrel is the fact that we have a "terror premium" factored into the price of a barrel of oil.

While it is important to consider the range of possibilities associated with a particular threat, assessments are typically based on the probability of a particular occurrence. Determining probability is somewhat subjective and is often based on intelligence and previous incidents. As mentioned, there are historical accounts of accidental incidents that have caused tremendous death and destruction.

Threat Warning Signs

A *threat warning* is an occurrence or discovery that indicates a potential threat that triggers an evaluation of the threat. It is important to note that these warnings must be evaluated in the context of typical industry activity and previous experience in order to avoid false alarms. Following is a brief description of potential warnings.

- Security Breach: Physical security breaches, such as unsecured doors, open hatches, or unlocked/forced gates, are probably the most common threat warnings. In most cases, the security breach is likely related to lax operations or typical criminal activity such as trespassing, vandalism, or theft. However, it may be prudent to assess any security breach with respect to the possibility of attack.
- Witness Account: Awareness of an incident may be triggered by a witness account of tampering. ESS sites/facilities should be aware that individuals observing suspicious behavior near ESS facilities will likely call 911 and not the facility. In this case, the incident warning technically might come from law enforcement, as described below. Note: the witness may be a plant employee engaged in their normal duties.
- Direct Notification by Perpetrator: A threat may be made directly to the manufacturing site, plant, or facility, either verbally or in writing. Historical incidents would indicate that verbal threats made over the phone are more likely than written threats. While the notification may be a hoax, threatening a manufacturing sector unit is a crime and should be taken seriously.

- Notification by Law Enforcement: A manufacturing site/facility may receive notification about a threat directly from law enforcement, including local, county, state, or federal agencies. As discussed previously, such a threat could be a result of suspicious activity reported to law enforcement, either by a perpetrator, a witness, or the news media. Other information, gathered through intelligence or informants, could also lead law enforcement to conclude that there may be a threat to the ESS site/facility. While law enforcement will have the lead in the criminal investigation, the ESS site/facility has the primary responsibility for the safety of its equipment and processes and for public health. Thus, the unit's role will likely be to help law enforcement to appreciate the public health implications of a particular threat as well as the technical feasibility of carrying out a particular threat.
- Notification by News Media: A threat to destroy an ESS site/facility might be delivered to the news media, or the media may discover a threat. A conscientious reporter would immediately report such a threat to the police, and either the reporter or the police would immediately contact the ESS site/facility. This level of professionalism would provide an opportunity for the plant to work with the media and law enforcement to assess the credibility of the threat before any broader notification is made.

RESPONSE TO THREATS

Note: This section is not designed to discuss what specific steps to take in responding to a terrorist threat. Rather, the questions addressed in this section are, Why is it necessary to plan to respond to ESS threats at all? and When have I done enough?

Federal, state, and local programs already exist that, with varying degrees of effectiveness, encourage or require the operators of ESS sites/facilities to boost their efforts to promote safety and security and to share information that can help local governments plan for emergencies.

Proper planning is a delicate process because public health measures are rarely noticed or appreciated (like buried utility pipes, they are often hidden functions) except when they fail—then they are very visible. Defense Department consumers are particularly upset by unreliable emergency services systems and associated services, or practices that produce unsafe (contaminated) environmental media—water, air, and/or soil—because they are often viewed as entitlements, and indeed it is reasonable for citizens to expect a high-quality, safe environment. Public health failures during a response to contamination threats often take the form of too much or too little action. The results of too little action, including no response at all, can have disastrous consequences, potentially resulting in public injuries or fatalities. Considering the potential risks of an inappropriate response to a severe ESS failure, it is clear that a systematic approach is needed to evaluate contamination threats. One overriding question is, When has an ESS unit done enough? This question may be particularly difficult to address when considering the wide range of agencies that may be involved in a threat situation. Other organizations, such as USFCC, USEPA, USDOE, CDC, USDOT, law-enforcement agencies, public health departments, etc., will each have unique obligations or interests in responding to a severe release or explosion threat.

When Is "Enough" Enough?

The guiding principle for responding to severe release or explosion threats is one of "due diligence" or "What is a suitable and sensible response to a threat?" As discussed above, some response to ESS failures is warranted due to the public health implications of an actual dangerous incident.

Ultimately, the answer to the question of due diligence must be decided at the local level and will depend on a number of considerations. Among other factors, local authorities must decide what level of risk is reasonable in the context of a perceived threat. Careful planning is essential to developing an appropriate response to terrorist threats, and in fact one primary objective of the Environmental Protection Agency's (EPA) Response Protocol Toolboxes (RPTBs) is to aid users in the development of their own site-specific plans that are consistent with the needs and responsibilities of the user. Beyond planning, the RPTB considers a careful evaluation of any terrorist threat, and an appropriate response based on the evaluation, to be the most important element of due diligence.

In the RPTB, the threat management process is considered in three successive stages: possible, credible, and confirmed. Thus, as the threat escalates through these three stages, the actions that might be considered due diligence expand accordingly. The following paragraphs describe, in general terms, actions that might be considered as due diligence at these various stages.

- Stage 1: "Is the threat possible?" If an ESS facility is faced with a terrorism threat, they should evaluate the available information to determine whether or not the threat is possible (i.e., could something have actually happened?). If the threat is possible, immediate operational response actions might be implemented, and activities such as site characterization would be initiated to collect additional information to support the next stage of the threat evaluation.

- Stage 2: "Is the threat credible?" Once a threat is considered possible, additional information will be necessary to determine if the threat is credible. The threshold at the credible stage is higher than that at the possible stage, and in general there must be information to corroborate the threat in order for it to be considered credible.
- Stage 3: "Has the incident been confirmed?" Confirmation implies that definitive evidence and information have been collected to establish the presence of a threat to the ESS. Obviously, at this stage the concept of due diligence takes on a whole new meaning since authorities are now faced with death and destruction and a potential public health crisis. Response actions at this point include all steps necessary to protect public health, property, and the environment.

PREPARATION

As an environmental/occupational safety, security, and health professional consultant for various utilities and other critical infrastructure sites on the East Coast, I have performed numerous security, pre-OSHA audit inspections and audits of various plant or facility process safety management/risk management planning (PSM/RMP) compliance programs. During these site visits, one factor always seemed to be universal. While conducting the plant/facility walk-around to gauge the organization's overall profile and status with FCC, OSHA, and EPA compliance, I almost always find that the plant/facility manager or superintendent who accompanies us is shocked to find out what is actually taking place at their facilities. They scratch their heads and ask various workers: What the hell are you doing? Where did that new computer come from? When was it installed? And also, why is that door broken? Who told you to paint that door, machine, or other apparatus? When did that hole get in the fence? Who left the back gate open? Where is the foreman? etc., etc., etc. Eventually, in an expression of utter consternation, the manufacturing manager/superintendent asks, "Who the hell is in charge around here?" And this is basically the question I find myself asking—far too often.

In one inspection performed at a plant/facility right after 9/11, I drove up to the entrance gate and was impressed with the height and condition of the barbed-razor-wire-topped fence and gates. I could not enter through the gates until I identified myself over a speaker system while a CCTV camera focused on my face. I was let in and given instructions to sign in at the main office. Not bad, just the way it should be . . . or so I thought at the time. After walking most of the plant site in the company of the plant manager, we approached the back fence area, which was close to a huge chlorine storage building. At the terminal end of the plant and fence, I noticed a large gate that was propped open with ivy growing on it, through it, and around it—obviously the gate had been in the open position for quite some time. I asked the plant manager why the gate was open. He stated that it was always open . . . that it led to a downhill path to a beach area below where plant personnel had constructed a picnic area, fronting the James River north of the Tidewater region of Virginia.

We walked the path to the bottom picnic area, looked around, and then looked back up the path toward the open gate and the prominent structure standing within, the chlorine storage building. While walking back up the path to the gate, I asked the plant manager if he was not concerned about the security and safety of the plant because the gate was left open, and especially about the safety of the fifty tons of deadly chlorine gas stored in the chlorine storage building.

"Nah . . . no way . . . we are safe here. I really don't see anyone swimming upriver just to get into the plant site. Ha! Besides, we are surrounded by woods out here . . . there's nothing to attack anyway."

Once inside the plant, I asked the plant manager if he was not worried about terrorists or some disgruntled former employee using a boat filled with explosives or some other weapon(s) gaining easy access to the plant and especially the chlorine building via the James River beach landing and picnic area.

"Nah, that will never happen . . . who would be that stupid? There's nothing around here worth blowin' up!"

Later, when I checked the geographic information system/Global Positioning System (GIS/GPS) data and maps pertaining to and showing the plant and surrounding area, I noted that about a half mile from the plant site was a large housing area, a brewery (with 700+ employees), and a very large theme and historical park—visited by more than 2.5 million people each summer.

Know Your Emergency Services Sector Systems

All ESS facility managers and manufacturing equipment operating personnel must know their plants/facilities. For these persons, there is no excuse for not knowing every square inch of the facility site. In particular, facility workers should know about any and all construction activities underway on the site, about the actual construction parameters of the plant, and especially about the operation of all plant unit processes. In addition, plant management must know not only their operating staff but also their customers.

Construction and Operation

Each ESS facility is unique with respect to age, operation, and complexity. This is important, particularly in evaluating the potential for a communications failure or malicious action(s).

Personnel

ESS employees are generally its most valuable asset in preparing for and responding to threats and incidents. They have knowledge of the system and potential problem areas. The importance of knowledgeable and experienced personnel is highlighted by the complexity of most catastrophes or other incidents. This complexity makes a specific terrorist target contingent upon detailed knowledge of the system configuration and usage patterns. If perpetrators have somehow gained a sophisticated understanding of an emergency services system, the day-to-day experience of network production will prove an invaluable tool to countering any attacks. For instance, personnel may continually look for unusual aspects of daily operation that might be interpreted as a potential threat warning, and they may also be aware of specific characteristics of the system that make it vulnerable to malware attacks or worse.

Perform Training and Desk/Field Exercises

In addition to a lack of planning, another reason that emergency response plans fail is lack of training and practice. Training provides the necessary means for everyone involved to acquire the skills to fulfill their role during an emergency. It may also provide important "buy-in" to the response process from both management and staff, which is essential to the success of any response plan. Desk exercise (also known as "tabletops," "sandlots," or "dry runs") along with field exercises allow participants to practice their skills. Also, these exercises will provide a test of the ESS security plan itself, revealing strengths and weaknesses that may be used to improve the overall plan.

Enhance Physical Security

Where possible, deny physical access to key sites; within the ESS system, this may act as a deterrent to a perpetrator. Terrorists often seek the easiest route of attack, just like a burglar prefers a house with an open window or an open automobile with keys in the ignition. Aside from deterring actual attacks, enhancing physical security has other benefits. For example, installation of fences and locks may reduce the rate of false alarms. Without surveillance equipment or locks, it may not be possible to determine whether a suspicious individual has actually entered a vulnerable area. The presence of a lock and a determination as to whether it has been cut or broken provides sound, although not definitive, evidence that an intrusion has occurred. Likewise, security cameras can be used to review security breaches and determine if the incident was simply due to trespassing or was a potential contamination threat. The cost of enhancing physical security may be justified by comparison to the cost of responding to just one "credible" munitions explosion or contamination threat involving site characterization and lab analysis for potential contaminants.

THE BOTTOM LINE

This chapter has emphasized the importance of ensuring the physical security of emergency services sector facilities and equipment. But it is important to point out that true ESS security goes beyond the physical facility; the sector requires security of a different kind. Much of the ESS is mobile and automated, including control architecture, and is vulnerable to cyberattack from either inside or outside the network. The fact is that perhaps the greatest vulnerability and dependency of the national ESS infrastructure as well as the infrastructure of other sectors is cybersecurity. The control of communications networks—essential for all critical infrastructure sectors—and all of its functional components are vulnerable to various degrees of cyberattack on the software operating systems, either by the idle hacker or by the more malicious intruder participating in information warfare. This is also true of all the networks in other sectors. The bottom line: the networks must be protected and guarded from attack.

REFERENCES AND RECOMMENDED READING

CBO. (2004). *Homeland Security and the Private Sector*. https://www.cbo.gov/sites/default/files/cbofiles/ftpdocs/60xx/doc6042/12-20-homelandsecurity.pdf, accessed May 17, 2023.

DHS. (2017). *Emergency Services Sector*. Washington, DC: Department of Homeland Security. https://www.dhs.gov/emergency-services-sector, accessed May 17, 2023.

Henry, K. (2002). "New Face of Security." *Government Security*, 30–37.

USEPA. 2001. *Protecting the Nation's Water Supplies from Terrorist Attack: Frequently Asked Questions*. Washington, DC: U.S. Environmental Protection Agency.

USEPA. 2004. *Response Protocol Toolbox: Planning for and Responding to Drinking Water Contamination Threats and Incidents*. Washington, DC: U.S. Environmental Protection Agency.

USEPA. 2011. *Response Protocol Toolbox: Planning for and Responding to Wastewater Contamination Threats and Incidents*. Washington, DC: U.S. Environmental Protection Agency.

Chapter 5

Cybersecurity and SCADA

Reports of cyber exploits illustrate the debilitating effects such attacks can have on the nation's security, economy, and public health and safety.

- In May 2015, media sources reported that data belonging to 1.1 million health insurance customers in the Washington, DC, area were stolen in a cyberattack on a private insurance company. Attackers accessed a database containing the names, birth dates, email addresses, and subscriber ID numbers of customers.
- In December 2014, the Industrial Control Systems Cyber Emergency Response Team (ICS-CERT; works to reduce risks within and across all critical infrastructure sectors by partnering with law-enforcement agencies) issued an updated alert on a sophisticated malware campaign compromising numerous industrial control system environments. Their analysis indicated that this campaign had been ongoing since at least 2011.
- In the January 2014–April 2014 release of its *Monitor* report, ICS-CERT reported that a public utility had been compromised when a sophisticated threat actor gained unauthorized access to its control system network through a vulnerable remote access capability configured on the system. The incident highlighted the need to evaluate security controls employed at the perimeter and to ensure that potential intrusion vectors are configured with appropriate security controls, monitoring, and detection capabilities.

In 2000, the Federal Bureau of Investigation (FBI) identified and listed threats to critical infrastructure. These threats are listed and described in table 5.1. In 2015, the U.S. Government Accountability Office (GAO) described the sources of cyber-based threats. These threats are listed and described in detail in table 5.2.

Table 5.1. Threats to Critical Infrastructure Observed by the FBI

Threat	*Description*
Criminal groups	There is an increased use of cyber intrusions by criminal groups who attack systems for purposes of monetary gain.
Foreign intelligence services	Foreign intelligence services use cyber tools as part of their information-gathering and espionage activities.
Hackers	Hackers sometimes crack into networks for the thrill of the challenge or for bragging rights in the hacker community. While remote cracking once required a fair amount of skill or computer knowledge, hackers can now download attack scripts and protocols from the internet and launch them against victim sites. Thus, while attack tools have become more sophisticated, they have also become easier to use.
Hacktivists	Hacktivism refers to politically motivated attacks on publicly accessible web pages or email servers. These groups and individuals overload email servers and hack into websites to send a political message.
Information warfare	Several nations are aggressively working to develop information warfare doctrine, programs, and capabilities. Such capabilities enable a single entity to have a significant and serious impact by disrupting the supply, communications, and economic infrastructures that support military power—impacts that, according to the director of central intelligence, can affect the daily lives of Americans across the country.
Inside threat	The disgruntled organization insider is a principal source of computer crimes. Insiders may not need a great deal of knowledge about computer intrusions because their knowledge of a victim system often allows them to gain unrestricted access to cause damage to the system or to steal system data. The insider threat also includes outsourcing vendors.
Virus writers	Virus writers are posing an increasingly serious threat. Several destructive computer viruses and "worms" have harmed files and hard drives, including the Melissa macro virus, the Explore.Zip worm, the CIH (Chernobyl) virus, Nimda, and Code Red.

Source: FBI (2000, 2014).

Table 5.2. Common Cyber-Threat Sources

Source	*Description*
Non-adversarial/Non-malicious	
Failure in information technology equipment responsible for data storage, processing, and communications	Failures in displays, sensors, controllers, and information
Failure in environmental controls	Failures in temperature/humidity controllers or power supplies
Software coding errors	Failures in operating systems, networking, and general-purpose and mission-specific applications
Natural or man-made disaster	Events beyond an entity's control such as fires, floods/tsunamis, tornadoes, hurricanes, and earthquakes
Unusual or natural event	Natural events beyond the entity's control that are not considered to be disasters (e.g., sunspots)
Infrastructure failure or outage	Failure or outage of telecommunications or electrical power
Unintentional user errors	Failures resulting from erroneous, accidental actions taken by individuals (both system users and administrators) in the course of executing their everyday responsibilities
Adversarial	
Hackers or hacktivists	Hackers break networks for the challenge, revenge, stalking, or monetary gain, among other reasons. Hacktivists are ideologically motivated actors who use cyber exploits to further political goals.
Malicious insiders	Insiders (e.g., disgruntled organization employees, including contractors) may not need a great deal of knowledge about computer intrusions because their position with the organization often allows them to gain unrestricted access and cause damage to the target system or to steal system data. These individuals engage in purely malicious activities and should not be confused with non-malicious insider accidents.
Nations	Nations, including nation-state, state-sponsored, and state-sanctioned programs, use cyber tools as part of their information-gathering and espionage activities. In addition, several nations are aggressively working to develop information warfare doctrine, programs, and capabilities.
Criminal groups and organized crime	Criminal groups seek to attack systems for monetary gain. Specifically, organized criminal groups use cyber exploits to commit identity theft, online fraud, and computer extortion.
Terrorist	Terrorists seek to destroy, incapacitate, or exploit critical infrastructures in order to threaten national security, cause mass casualties, weaken the economy, and damage public morale and confidence.
Unknown malicious outsiders	Unknown malicious outsiders are threat sources or agents that, due to a lack of information, agencies are unable to classify as being one of the five types of threat sources or agents listed above.

Source: GAO analysis of unclassified government and nongovernmental data. GAO (2015), pp. 16–79.

Threats to systems supporting critical infrastructure are evolving and growing. As shown in table 5.2, cyber threats can be unintentional or intentional. Unintentional or non-adversarial threats include equipment failures, software coding errors, and the actions of poorly trained employees. They also include natural disasters and failures of critical infrastructure on which the organization depends but are outside of its control. Intentional threats include both targeted and untargeted attacks from a variety of sources, including criminal groups, hackers, disgruntled employees, foreign nations engaged in espionage or information warfare, and terrorists. These threat adversaries vary in terms of the capabilities of the actors, their willingness to act, and their motives, which can include seeking monetary gain or seeking an economic, political, or military advantage (GAO, 2015).

CYBERSPACE

Today's developing "information age" technology has intensified the importance of critical infrastructure protection, in which cybersecurity has become as critical as physical security to protecting virtually all critical infrastructure sectors. The Department of Defense (DOD) has determined that cyber threats to contractors' unclassified information systems represent an unacceptable risk of compromise of DOD information and pose a significant risk to U.S. national security and economic security interests.

In the past few years, especially since 9/11, it has been somewhat routine for us to pick up a newspaper or magazine or view a television news program where a major topic of discussion is cybersecurity or the lack thereof. For example, recently there has been discussion about Russian hackers trying to influence the 2016 U.S. elections. Many of the cyber-intrusion incidents we read or hear about have added new terms, or new uses for old terms, to our vocabulary. For example, old terms such as *botnets* (short for "robot networks," also called *bots*, *zombies*, *botnet fleets*, and many others), which refers to groups of computers that have been compromised with malware such as Trojan horses, worms, back doors, remote-control software, and viruses, have taken on new connotations in regard to cybersecurity issues. Relatively new terms such as *scanners*, *Windows NT hacking tools*, *ICQ hacking tools*, *mail bombs*, *sniffers*, *logic bombs*, *nukers*, *dots*, *backdoor Trojans*, *keyloggers*, *hackers' Swiss knives*, *password crackers*, *blended threats*, *Warhol worms*, *Flash threats*, *targeted attacks*, and *BIOS crackers* are now commonly read or heard. New terms have evolved along with various control mechanisms. For example, because many control systems are vulnerable to attacks of varying degrees, these attack attempts range from telephone line sweeps (war dialing), to wireless network sniffing (war driving), to physical network port scanning, to physical monitoring and intrusion. When wireless network sniffing is performed at (or near) the target point by a pedestrian, it is called war walking, meaning that instead of a person being in an automotive vehicle, the potential intruder may be sniffing the network for weaknesses or vulnerabilities on foot, posing as a person walking, but they may have a handheld PDA device or laptop computer (Warwalking 2003). Further, adversaries can leverage common computer software programs, such as Adobe Acrobat or Microsoft Office, to deliver a threat by embedding exploits within software files that are activated when a user opens a file within its corresponding program. Finally, ESS infrastructure is extremely dependent on the IT sector. This dependency is due to the reliance of communications systems on the software that runs the control mechanism of the operations systems, the management software, the machining software, and any number of other software packages used by industry. Table 5.3 provides descriptions of common exploits or techniques, tactics, and practices used by cyber adversaries.

Not all relatively new and universally recognizable cyber terms have sinister connotations or meanings, of course. Consider, for example, the following digital terms: *backup*, *binary*, *bit*, *byte*, *CD-ROM*, *CPU*, *database*, *email*, *HTML*, *icon*, *memory*, *cyberspace*, *modem*, *monitor*, *network*, *RAM*, *Wi-Fi* (wireless fidelity), *record*, *software*, *World Wide Web*—none of these terms normally generate thoughts of terrorism in most of us.

There is, however, one digital term, *SCADA*, that most people have not heard of. This is not the case, however, for those who work with the nation's critical infrastructure, including ESS infrastructure. SCADA, or *supervisory control and data acquisition*, systems (also sometimes referred to as "digital control systems" or "process control systems") provide the real-time control mechanisms for most power and utility facilities and play an important role in computer-based control systems. From coordinating music and lights in proper sequence and the vaulting of spray from water fountains, to controlling systems used in the drilling and refining of oil and natural gas, control systems perform many functions. Many energy distribution networks use computer-based systems to remotely control sensitive feeds, processes, and system equipment previously controlled manually. SCADA systems allow an energy utility (or operation) to monitor fuel tank levels, ensure that contents are stored at correct levels, and, as mentioned, collect data from sensors and control equipment located at remote sites. Common communications sector and customer system sensors measure many parameters, depending on the nature of the industrial manufacturing entity. Common SCADA customer industry system equipment includes valves, pumps, and switching devices for distribution of electricity. The critical infrastructure of many countries is increasingly dependent on SCADA systems.

Table 5.3 Common Methods of Cyber Exploits

Exploit	*Description*
Watering hole	A method by which threat actors exploit the vulnerabilities of carefully selected websites frequented by users of the targeted system. Malware is then injected into the targeted system via the compromised websites.
Phishing and spear phishing	A digital form of social engineering that uses authentic-looking emails, websites, or instant messages to get users to download malware, open malicious attachments, or open links that direct then to a website that requires information or executes malicious code.
Credentials based	An exploit that takes advantage of a system's insufficient user authentication and/or any elements of cybersecurity supporting it, including not limiting the number of failed login attempts, the use of hard-coded credentials, and the use of a broken or risky cryptographic algorithm.
Trusted third parties	An exploit that takes advantage of the security vulnerabilities of trusted third parties to gain access to an otherwise secure system.
Classic buffer overflow	An exploit that involves the intentional transmission of more data than a program's input buffer can hold, leading to the deletion of critical data and subsequent execution of malicious code.
Cryptographic weakness	An exploit that takes advantage of a network employing insufficient encryption when either storing or transmitting data, enabling adversaries to read and/or modify the data stream.
Structured Query Language (SQL) injection	An exploit that involves the alteration of a database search in a web-based application, which can be used to obtain unauthorized access to sensitive information in a database, resulting in data loss or corruption, denial or service, or complete host takeover.
Operating system command injection	An exploit that takes advantage of a system's inability to properly neutralize special elements used in operating system commands, allowing the adversaries to execute unexpected commands on the system by either modifying already evoked commands or evoking their own.
Cross-site scripting	An exploit that uses third-party web resources to run lines of programming code (referred to as scripts) within the victim's web browser or scriptable application. This occurs when a user, using a browser, visits a malicious website or clicks a malicious link. The most dangerous consequences can occur when this method is used to exploit additional vulnerabilities that may permit an adversary to steal cookies (data exchanged between a web server and a browser), log keystrokes, capture screenshots, discover and collect network information, or remotely access and control the victim's machine.

Exploit	*Description*
Cross-site request forgery	An exploit takes advantage of an application that cannot, or does not, sufficiently verify whether a well-formed, valid, and consistent request was intentionally provided by the user who submitted the request, tricking the victim into executing a falsified request that results in the system or data being compromised.
Path traversal	An exploit that seeks to gain access to files outside of a restricted directory by modifying the directory path name in an application that does not properly neutralize special elements (e.g., "...", "/", ".../").
Integer overflow	An exploit where malicious code is inserted that leads to unexpected integer overflow, or wraparound, which can be used by adversaries to control looping or make security decisions in order to cause program crashes, memory corruption, or the execution of arbitrary code via buffer overflow.
Uncontrolled format string	Adversaries manipulate externally controlled format strings in print-style functions to gain access to information and/or execute unauthorized code or commands.
Open redirect	An exploit where the victim is tricked into selecting a URL (website location) that has been modified to direct them to an external, malicious site which may contain malware that can compromise the victim's machine.
Heap-based buffer overflow	Similar to classic buffer overflow, but the buffer that is overwritten is allocated in the heap portion of memory, generally meaning that the buffer was allocated using a memory allocation routine, such as "malloc ( )."
Unrestricted upload of files	An exploit that takes advantage of insufficient upload restrictions, enabling adversaries to upload malware (e.g., .php) in place of the intended file type (e.g., .jpg).
Inclusion of functionality from untrusted sphere	An exploit that uses trusted third-party executable functionality (e.g., web widget or library) as a means of executing malicious code in software whose protection mechanisms are unable to determine whether functionality is from a trusted source, was modified in transit, or is being spoofed.
Certificate and certificate authority compromise	Exploits facilitated via the issuance of fraudulent digital certificates (e.g., transport layer security and Secure Sockets Layer). Adversaries use these certificates to establish secure connections with the target organization or individual by mimicking a trusted third party.
Hybrid of others	An exploit that combines elements of two or more of the aforementioned techniques.

Source: GAO (2015).

WHAT IS SCADA?

If we were to ask the specialist to define SCADA, the technical response could be outlined as follows:

- It is a multitier system or interfaces with multitier systems.
- It is used for physical measurement and control endpoints via a remote terminal unit (RTU) and programmable logic controller (PLC) to measure voltage, adjust a value, or flip a switch.
- It is an intermediate processor normally based on commercial third-party operating systems—VMS, Unix, Windows, Linux.
- It human interfaces, for example, with a graphical user interface (Windows GUIs).
- Its communication infrastructure consists of a variety of transport mediums such as analog, serial, internet, radio, or Wi-Fi.

How about the nonspecialist response?—for the rest of us who are nonspecialists. Well, for those of us in this category, we could simply say that SCADA is a computer-based control system that remotely controls and monitors processes previously controlled and monitored manually. The philosophy behind SCADA control systems can be summed up by the phrase "If you can measure it, you can control it." SCADA allows an operator using a central computer to supervise (control and monitor) multiple networked computers at remote locations. Each remote computer can control mechanical processes (machines, mixers, pumps, valves, etc.) and collect data from sensors at a remote location. Thus the phrase: supervisory control and data acquisition, or SCADA.

The central computer is called the master terminal unit, or MTU. The MTU has two main functions: periodically obtain data from RTUs/PLCs and control remote devices through the operator station. The operator interfaces with the MTU using software called a human-machine interface, or HMI. The remote computer is called a programmable logic controller (PLC) or remote terminal unit (RTU). The RTU activates a relay (or switch) that turns mechanical equipment "on" and "off." The RTU also collects data from sensors. Sensors perform measurements, and actuators perform control.

In the initial stages of this technology, utilities ran wires, also known as hardwire or landlines, from the central computer (MTU) to the remote computers (RTUs). Because remote locations can be located hundreds of miles from the central location, utilities began to use public phone lines and modems, leased telephone company lines, and radio wave or microwave communication. More recently, they have also begun to use satellite links, the internet, and newly developed wireless technologies.

Because SCADA systems' sensors provide valuable information, many critical infrastructure entities, utilities, and other industries established "con-

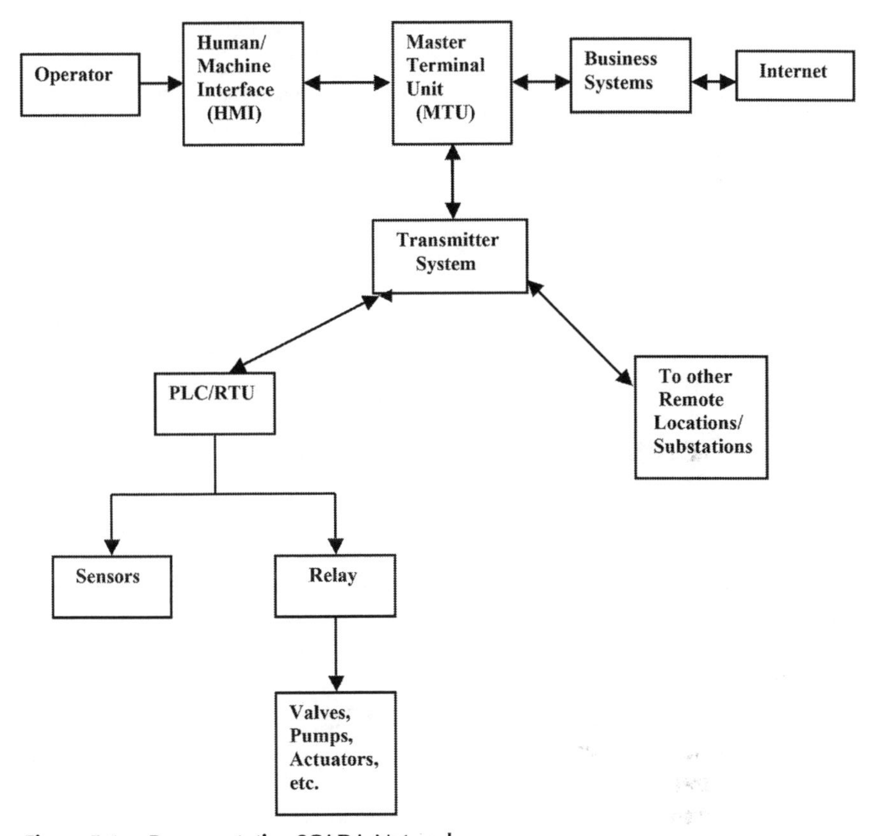

Figure 5.1. Representative SCADA Network

nections" between their SCADA systems and their business systems. This allowed utility/industrial management and other staff access to valuable statistics, such as chemical usage. When utilities/industries later connected their systems to the internet, they were able to provide stakeholders/stockholders with usage statistics on the emergency services segment, utility/industrial, and some emergency services web pages. Figure 5.1 provides a basic illustration of a representative SCADA network. Note that firewall protection would normally be placed between the internet and the business system and between the business system and the MTU.

SCADA APPLICATIONS

As stated above, SCADA systems can be designed to measure a variety of equipment operating conditions and parameters, volumes and flow rates, and electricity, natural gas, oil, and petrochemical mixture quality parameters, and to respond to changes in those parameters, either by alerting operators or by modifying system operation through a feedback-loop system, without personnel having to physically visit each valve, process, or piece of other equipment on a daily basis to check it to ensure that it is functioning properly. Automation and integration of large-scale diverse assets required SCADA systems to provide the utmost in flexibility, scalability, openness, and reliability. SCADA systems are used to automate certain energy-production functions; these can be performed without initiation by an operator. In addition to process equipment, SCADA systems can also integrate specific security alarms and equipment, such as cameras, motion sensors, lights, data from card-reading systems, etc., thereby providing a clear picture of what is happening at areas throughout a facility. Finally, SCADA systems also provide constant, real-time data on processes, equipment, location access, etc., allowing the necessary response to be made quickly. This can be extremely useful during emergency conditions, such as when energy distribution lines or piping breaks or when potentially disruptive chemical reaction spikes appear in chemical processing operations. Currently, it can be said that SCADA has evolved from a simple indicating light and push-button control system into a comprehensive operation and handling system for very complex process control and safety shutdown systems. In a nutshell, SCADA results in an oversight system that requires fewer operators.

Today, many common ESS applications for SCADA systems include those highlighted below, but are not limited to these:

- Boiler controls
- Bearing temperature monitors (in electric generators and motors)
- Machine operation
- Gas processing
- Plant monitoring
- Plant energy management
- Power distribution monitoring
- Electric power monitoring
- Fuel oil handling system
- Plant monitoring
- Process controls
- Process stimulators
- Tank controls
- Utility monitoring
- Safety parameter display systems and shutdown systems
- Tank level control and monitoring
- Turbine controls

- Turbine monitoring
- Virtual annunciator panels
- Alarm systems (can play a key role in ESS)
- Security equipment (ESS related)
- Event logging (ESS related)

Because these systems can monitor multiple processes, equipment, and infrastructure and then provide quick notification of, or response to, problems or upsets, SCADA systems typically provide the first line of detection for atypical or abnormal conditions. For example, a SCADA system may be connected to sensors that measure specific machining quality parameters that are outside of a specific range. A real-time customized operator interface screen could display and control critical systems monitoring these parameters.

The system could transmit warning signals back to the operators, such as by initiating a call to a personal pager. This might allow the operators to initiate actions to prevent power outages or contamination. Further automation of the system could ensure that the system initiated measures to rectify the problem. Preprogrammed control functions (e.g., shutting a valve, controlling flow, throwing a switch, or adding chemicals) can be triggered and operated based on SCADA utility.

SCADA VULNERABILITIES

U.S. Electric Grid Gets Hacked Into

The Associated Press (AP) reported on April 9, 2009, that spies hacked into the U.S. energy grid and left behind computer programs (Trojan horses) that would enable them to disrupt service, exposing potentially catastrophic vulnerabilities in key pieces of national infrastructure.

Even though terrorists, domestic and foreign, tend to aim their main focus around the critical devices that control actual communications system delivery activities, according to USEPA (2005), SCADA networks were developed with little attention paid to security, often making the security of these systems weak. Studies have found that, while technological advancements introduced vulnerabilities, many critical infrastructure facilities and utilities have spent little time securing their SCADA networks. As a result, many SCADA networks may be susceptible to attacks and misuse. SCADA systems languished in obscurity, and this was the essence of their security—that is, until technological developments transformed SCADA from a backroom operation to a front-and-center visible control system.

Remote monitoring and supervisory control of processes began to develop in the early 1960s and adopted many technological advancements. The advent of minicomputers made it possible to automate a vast number of once manually operated switches. Advancements in radio technology reduced the communication costs associated with installing and maintaining buried cable in remote areas. SCADA systems continued to adopt new communication methods, including satellite and cellular. As the price of computers and communications dropped, it became economically feasible to distribute operations and to expand SCADA networks to include even smaller facilities.

Advances in information technology and the necessity of improved efficiency have resulted in increasingly automated and interlinked infrastructures and created new vulnerabilities due to equipment failure, human error, weather and other natural causes, and physical and cyber attacks. Some areas and examples of possible SCADA vulnerabilities include the following (Wiles et al., 2007):

- Human: People can be tricked or corrupted, and may commit errors.
- Communications: Messages can be fabricated, intercepted, changed, deleted, or blocked.
- Hardware: Security features are not easily adapted to small self-contained units with limited power supplies.
- Physical: Intruders can break into a facility to steal or damage SCADA equipment.
- Natural: Tornadoes, floods, earthquakes, and other natural disasters can damage equipment and connections.
- Software: Programs can be poorly written.

Specific SCADA weaknesses and potential attack vectors include the following:

- Does not require any authorization
- Does not use encryption
- Does not properly recognize and properly handle errors and exceptions
- No authentication is required
- Data can be intercepted
- Manipulation of data
- Service denial
- IP address spoofing—internet protocol packets with a false source IP address
- Session hijacking
- Unsolicited responses
- Packet fuzzing—involves inputting false information, etc.

- Unauthorized control
- Log data manipulation

SCADA system computers and their connections are susceptible to different types of information system attacks and misuse, such as those mentioned above. The Computer Security Institute and the Federal Bureau of Investigation conduct an annual Computer Crime and Security Survey (FBI, 2007). The 2007 survey reported on ten types of attack or misuse and found that viruses and denial of service had the greatest negative economic impact. The same study also found that 15 percent of the respondents reported abuse of wireless networks, which can be a SCADA component. On average, respondents from all sectors did not believe that their organization invested enough in security awareness. For example, utilities as a group reported a lower average computer security expenditure/investment per employee than many other sectors such as transportation, telecommunications, and finance.

Sandia National Laboratories' *Common Vulnerabilities in Critical Infrastructure Control Systems* described some of the common problems it has identified in the following five categories (Stamp et al., 2003):

1. System Data: Important data attributes for security include availability, authenticity, integrity, and confidentiality. Data should be categorized according to its sensitivity, and ownership and responsibility must be assigned. However, SCADA data is often not classified at all, making it difficult to identify where security precautions are appropriate (for example, which communication links to secure databases require protection).
2. Security Administration: Vulnerabilities emerge because many systems lack a properly structured security policy (security administration is notoriously lax in the case of control systems), equipment and system implementation guides, configuration management, training, and enforcement and compliance auditing.
3. Architecture: Many common practices negatively affect SCADA security. For example, while it is convenient to use SCADA capabilities for other purposes such as fire and security systems, these practices create single points of failure. Also, the connection of SCADA networks to other automation systems and business networks introduces multiple entry points for potential adversaries.
4. Network (including communication links): Legacy systems' hardware and software have very limited security capabilities, and the vulnerabilities of contemporary systems (based on modern information technology) are publicized. Wireless and shared links are susceptible to eavesdropping and data manipulation.

5. Platforms: Many platform vulnerabilities exist, including default configurations retained, poor password practices, shared accounts, inadequate protection for hardware, and nonexistent security monitoring controls. In most cases, important security patches are not installed, often due to concern about negatively impacting system operation; in some cases technicians are contractually forbidden from updating systems by their vendor agreements.

The following incident helps to illustrate some of the risks associated with SCADA vulnerabilities: During the course of conducting a vulnerability assessment, a contractor stated that personnel from his company penetrated the information system of a utility within minutes. Contractor personnel drove to a remote substation and noticed a wireless network antenna. Without leaving their vehicle, they plugged in their wireless radios and connected to the network within five minutes. Within twenty minutes they had mapped the network, including SCADA equipment, and accessed the business network and data. This illustrates what a cybersecurity advisor from Sandia National Laboratories specializing in SCADA stated, that utilities are moving to wireless communication without understanding the added risks.

The Increasing Risk

According to the GAO (2015), historically, security concerns about control systems (SCADA included) were related primarily to protecting against physical attack and misuse of refining and processing sites or distribution and holding facilities. However, more recently there has been a growing recognition that control systems are now vulnerable to cyberattacks from numerous sources, including hostile governments, terrorist groups, disgruntled employees, and other malicious intruders.

In addition to the control system vulnerabilities mentioned earlier, several factors have contributed to the escalation of risks to control systems, including (1) the adoption of standardized technologies with known vulnerabilities, (2) the connectivity of control systems to other networks, (3) constraints on the implementation of existing security technologies and practices, (4) insecure remote connections, and (5) the widespread availability of technical information about control systems.

Adoption of Technologies with Known Vulnerabilities

When a technology is not well known, not widely used, or not understood or publicized, it is difficult to penetrate it and thus disable it. Historically, proprietary hardware, software, and network protocols made it difficult to understand how control systems operated—and therefore how to hack into them. Today, however, to reduce costs and improve performance, organizations have been transitioning from proprietary systems to less expensive, standardized technologies such as Microsoft's Windows and Unix-like operating systems and the common networking protocols used by the internet. These widely used standardized technologies have commonly known vulnerabilities, and sophisticated and effective exploitation tools are widely available and relatively easy to use. As a consequence, both the number of people with the knowledge to wage attacks on these systems and the number of systems subject to attack have increased. Also, common communication protocols and the emerging use of Extensible Markup Language (commonly referred to as XML) can make it easier for a hacker to interpret the content of communications among the components of a control system.

Control systems are often connected to other networks—enterprises often integrate their control system with their enterprise networks. This increased connectivity has significant advantages, including providing decision-makers with access to real-time information and allowing engineers to monitor and control the process control system from different points on the enterprise network. In addition, the enterprise networks are often connected to the networks of strategic partners and to the internet. Further, control systems are increasingly using wide area networks and the internet to transmit data to their remote or local stations and individual devices. This convergence of control networks with public and enterprise networks potentially exposes the control systems to additional security vulnerabilities. Unless appropriate security controls are deployed in the enterprise network and the control system network, breaches in enterprise security can affect the operation of the control system.

According to industry experts, the use of existing security technologies, as well as strong user authentication and patch management practices, are generally not implemented in control systems because control systems operate in real time, typically are not designed with cybersecurity in mind, and usually have limited processing capabilities.

Existing security technologies such as authorization, authentication, encryption, intrusion detection, and filtering of network traffic and communications require more bandwidth, processing power, and memory than control system components typically have. Because controller stations are generally designed to do specific tasks, they use low-cost, resource-constrained microprocessors. In fact, some devices in the electrical industry still use the Intel 8088 processor, introduced in 1978. Consequently, it is difficult to install existing security technologies without seriously degrading the performance of the control system.

Further, complex passwords and other strong password practices are not always used to prevent unauthorized access to control systems, in part because this could hinder a rapid response to safety procedures during an emergency. As a result, according to experts, weak passwords that are easy to guess, shared, and infrequently changed are reportedly common in control systems, including the use of default passwords or even no password at all.

In addition, although modern control systems are based on standard operating systems, they are typically customized to support control system applications. Consequently, vendor-provided software patches are generally either incompatible or cannot be implemented without compromising service, shutting down "always-on" systems, or affecting interdependent operations.

Potential vulnerabilities in control systems are exacerbated by insecure connections. Organizations often leave access links—such as dial-up modems to equipment and control information—open for remote diagnostics, maintenance, and examination of system status. Such links may not be protected with authentication or encryption, which increases the risk that hackers could use these insecure connections to break into remotely controlled systems. Also, control systems often use wireless communication systems, which are especially vulnerable to attack, or leased lines that pass through commercial telecommunications facilities. Without encryption to protect data as it flows through these insecure connections or authentication mechanisms to limit access, there is limited protection for the integrity of the information being transmitted.

Public information about infrastructures and control systems is available to potential hackers and intruders. The availability of this infrastructure and vulnerability data was demonstrated by a university graduate student whose dissertation reportedly mapped every business and industrial sector in the American economy to the fiber-optic network that connects them—using material that was available publicly on the internet, none of which was classified.

Cyber Threats to Control Systems

There is a general consensus—and increasing concern—among government officials and experts on control systems about potential cyber threats to the control systems that govern our critical infrastructures. As components of control systems increasingly make critical decisions that were once made by humans, the potential effect of a cyber threat becomes more devastating. Such cyber threats could come from numerous sources, ranging from hostile governments and terrorist groups to disgruntled employees and other malicious intruders.

In July 2002, the National Infrastructure Protection Center (NIPC) reported that the potential for compound cyber and physical attacks, referred to as "swarming attacks," is an emerging threat to U.S. critical infrastructure. As the NIPC reports, the effects of a swarming attack include slowing or complicating the response to a physical attack. For instance, a cyberattack that disabled the water supply or the electrical system in conjunction with a physical attack could deny emergency services the necessary resources to manage the consequences of the attack—such as controlling fires, coordinating actions, and generating light.

Control systems such as SCADA can be vulnerable to cyberattacks. Entities or individuals with malicious intent might take one or more of the following actions to successfully attack control systems:

- Disrupt the operation of control systems by delaying or blocking the flow of information through control networks, thereby denying availability of the networks to control system operations.
- Make unauthorized changes to programmed instructions in PLCs, RTUs, or distributed control system (DCS) controllers; change alarm thresholds; or issue unauthorized commands to control equipment, which could potentially result in damage to equipment (if tolerances are exceeded), premature shutdown of processes (such as prematurely shutting down transmission lines), or even the disabling of control equipment.
- Send false information to control system operators either to disguise unauthorized changes or to initiate inappropriate actions by system operators.
- Modify the control system software, producing unpredictable results.
- Interfere with the operation of safety systems.

In addition, in control systems that cover a wide geographic area, the remote sites are often unstaffed and may not be physically monitored. If such remote systems are physically breached, the attackers could establish a cyber connection to the control network.

Securing Control Systems

Several challenges must be addressed to effectively secure control systems against cyber threats.

These challenges include (1) the limitations of current security technologies in securing control systems, (2) the perception that securing control systems may not be economically justifiable, and (3) the conflicting priorities within organizations regarding the security of control systems.

A significant challenge in effectively securing control systems is the lack of specialized security technologies for these systems. The computing resources in control systems that are needed to perform security functions tend to be quite limited, making it very difficult to use security technologies within control system networks without severely hindering performance.

Securing control systems may not be perceived as economically justifiable. Experts and industry representatives have indicated that organizations may be reluctant to spend more money to secure control systems. Hardening the security of control systems would require industries to expend more resources, including acquiring more personnel, providing training for personnel, and potentially prematurely replacing current systems that typically have a lifespan of about twenty years.

Finally, several experts and industry representatives have indicated that the responsibility for securing control systems typically includes two separate groups: IT security personnel and control system engineers and operators. IT security personnel tend to focus on securing enterprise systems, while control system engineers and operators tend to be more concerned with the reliable performance of their control systems. Further, they indicate that, as a result, these two groups do not always fully understand each other's requirements and thus do not collaborate to implement secure control systems.

STEPS TO IMPROVING SCADA SECURITY

The President's Critical Infrastructure Protection Board and the Department of Energy (DOE) have developed the steps outlined below to help organizations improve the security of their SCADA networks. The DOE (2001) points out that these steps are not meant to be prescriptive or all inclusive. However, they do address essential actions to be taken to improve the protection of SCADA networks. The steps are divided into two categories: specific actions to improve implementation and actions to establish essential underlying management processes and policies.

Twenty-One Steps to Increase SCADA Security (DOE, 2001)

The following steps focus on specific actions to be taken to increase the security of SCADA networks:

1. Identify all connections to SCADA networks.
   Conduct a thorough risk analysis to assess the risk and necessity of each connection to the SCADA network. Develop a comprehensive under-

standing of all connections to the SCADA network and how well those connections are protected. Identify and evaluate the following types of connections:

- Internal local area and wide area networks, including business networks
- The internet
- Wireless network devices, including satellite uplinks
- Modem or dial-up connections
- Connections to business partners, vendors, or regulatory agencies

2. Disconnect unnecessary connections to the SCADA network.
   To ensure the highest degree of security of SCADA systems, isolate the SCADA network from other network connections to as great a degree as possible. Any connection to another network introduces security risks, particularly if the connection creates a pathway from or to the internet. Although direct connections with other networks may allow important information to be passed efficiently and conveniently, insecure connections are simply not worth the risk; isolation of the SCADA network must be a primary goal to provide needed protection. Strategies such as utilization of "demilitarized zones" (DMZs) and data warehousing can facilitate the secure transfer of data from the SCADA network to business networks. However, they must be designed and implemented properly to avoid introduction of additional risk through improper configuration.

3. Evaluate and strengthen the security of any remaining connections to the SCADA networks.
   Conduct penetration testing or vulnerability analysis of any remaining connections to the SCADA network to evaluate the protection posture associated with these pathways. Use this information in conjunction with risk management processes to develop a robust protection strategy for any pathways to the SCADA network. Since the SCADA network is only as secure as its weakest connecting point, it is essential to implement firewalls, intrusion detection systems (IDSs), and other appropriate security measures at each point of entry. Configure firewall rules to prohibit access from and to the SCADA network, and be as specific as possible when permitting approved connections. For example, an independent system operator (ISO) should not be granted "blanket" network access simply because there is a need for a connection to certain components of the SCADA system. Strategically place IDSs at each entry point to alert security personnel of potential breaches of network security. Organization management must understand and accept responsibility for risks associated with any connection to the SCADA network.

4. Harden SCADA networks by removing or disabling unnecessary services. SCADA control servers built on commercial or open-source operating systems can be exposed to attack through default network services. To the greatest degree possible, remove or disable unused services and network demons to reduce the risk of direct attack. This is particularly important when SCADA networks are interconnected with other networks. Do not permit a service or feature on a SCADA network unless a thorough risk assessment of the consequences of allowing the service/feature shows that the benefits of the service/feature far outweigh the potential for vulnerability exploitation. Examples of services to remove from SCADA networks include automated meter reading/remote billing systems, email services, and internet access. An example of a feature to disable is remote maintenance. Utilize numerous secure configurations such as the National Security Agency's series of security guides. Additionally, work closely with SCADA vendors to identify secure configurations and coordinate any and all changes to operational systems to ensure that removing or disabling services does not cause downtime, interruption of service, or loss of support.

5. Do not rely on proprietary protocols to protect your system.
   Some SCADA systems are unique, proprietary protocols for communications between field devices and servers. Often the security of SCADA systems is based solely on the secrecy of these protocols. Unfortunately, obscure protocols provide very little "real" security. Do not rely on proprietary protocols or factory default configuration settings to protect your system. Additionally, demand that vendors disclose any back doors or vendor interfaces to your SCADA systems, and expect them to provide systems that are capable of being secured.

6. Implement the security features provided by device and system vendors.
   Older SCADA systems (most systems in use) have no security features whatsoever. SCADA system owners must insist that their system vendor implement security features in the form of product patches or upgrades. Some newer SCADA devices are shipped with basic security features, but these are usually disabled to ensure ease of installation.

   Analyze each SCADA device to determine whether security features are present. Additionally, factory default security settings (such as in computer network firewalls) are often set to provide maximum usability but minimal security. Set all security features to provide the maximum security only after a thorough risk assessment of the consequences of reducing the security level.

7. Establish strong controls over any medium that is used as a back door into the SCADA network.
   Where back doors or vendor connections do exist in SCADA systems, strong authentication must be implemented to ensure secure communications. Modems, wireless, and wired networks used for communications and maintenance represent a significant vulnerability to the SCADA network and remote sites. Successful "war dialing" or "war driving" attacks could allow an attacker to bypass all other controls and have direct access to the SCADA network or resources. To minimize the risk of such attacks, disable inbound access and replace it with some type of callback system.

8. Implement internal and external intrusion detection systems and establish twenty-four-hour-a-day incident monitoring.
   To be able to effectively respond to cyberattacks, establish an intrusion detection strategy that includes alerting network administrators of malicious network activity originating from internal or external sources. Intrusion detection system monitoring is essential twenty-four hours a day; this capability can be easily set up through a pager. Additionally, incident response procedures must be in place to allow an effective response to any attack. To complement network monitoring, enable logging on all systems and audit system logs daily to detect suspicious activity as soon as possible.

9. Perform technical audits of SCADA devices and networks, and any other connected networks, to identify security concerns.
   Technical audits of SCADA devices and networks are critical to ongoing security effectiveness. Many commercial and open-source security tools are available that allow system administrators to conduct audits of their systems/networks to identify active services, patch level, and common vulnerabilities. The use of these tools will not solve systemic problems but will eliminate the "paths of least resistance" that an attacker could exploit. Analyze identified vulnerabilities to determine their significance and take corrective actions as appropriate. Track corrective actions and analyze this information to identify trends. Additionally, retest systems after corrective actions have been taken to ensure that vulnerabilities were actually eliminated. Scan nonproduction environments actively to identify and address potential problems.

10. Conduct physical security surveys and assess all remote sites connected to the SCADA network to evaluate their security.
    Any location that has a connection to the SCADA network is a target, especially unmanned or unguarded remote sites. Conduct a physical security survey and inventory access points at each facility that has a

connection to the SCADA system. Identify and assess any source of information, including remote telephone/computer network/fiber-optic cables, that could be tapped; radio and microwave links that are exploitable; computer terminals that could be accessed; and wireless local area network access points. Identify and eliminate single points of failure. The security of the site must be adequate to detect or prevent unauthorized access. Do not allow "live" network access points at remote, unguarded sites simply for convenience.

11. Establish SCADA "red teams" to identify and evaluate possible attack scenarios.
    Establish a "red team" to identify potential attack scenarios and evaluate potential system vulnerabilities. Use a variety of people who can provide insight into weaknesses of the overall network, SCADA system, physical systems, and security controls. People who work on the system every day have great insight into the vulnerabilities of your SCADA network and should be consulted when identifying potential attack scenarios and possible consequences. Also, ensure that the risk from a malicious insider is fully evaluated, given that this represents one of the greatest threats to an organization. Feed information resulting from the "red team" evaluation into risk management processes to assess the information and establish appropriate protection strategies.

The following steps focus on management actions to establish an effective cybersecurity program:

12. Clearly define cybersecurity roles, responsibilities, and authorities for managers, system administrators, and users.
    Organization personnel need to understand the specific expectations associated with protecting IT resources through the definition of clear and logical roles and responsibilities. In addition, key personnel need to be given sufficient authority to carry out their assigned responsibilities. Too often, good cybersecurity is left up to the initiative of the individual, which usually leads to inconsistent implementations and ineffective security. Establish a cybersecurity organizational structure that defines roles and responsibilities and clearly identifies how cybersecurity issues are escalated and who is notified in an emergency.

13. Document network architecture and identify systems that serve critical functions or contain sensitive information that require additional levels of protection.

Develop and document robust information security architecture as part of a process to establish an effective protection strategy. It is essential that organizations design their network with security in mind and continue to have a strong understanding of their network architecture throughout its life cycle. Of particular importance, an in-depth understanding of the functions that the systems perform and the sensitivity of the stored information is required. Without this understanding, risk cannot be properly assessed, and protection strategies may not be sufficient. Documenting the information security architecture and its components is critical to understanding the overall protection strategy and identifying single points of failure.

14. Establish a rigorous, ongoing risk management process.
    A thorough understanding of the risks to network computing resources from denial-of-service attacks and the vulnerability of sensitive information to compromise is essential to an effective cybersecurity program. Risk assessments form the technical basis of this understanding and are critical to formulating effective strategies to mitigate vulnerabilities and preserve the integrity of computing resources. Initially, perform a baseline risk analysis based on current threat assessment to use for developing a network protection strategy. Due to rapidly changing technology and the emergence of new threats on a daily basis, an ongoing risk assessment process is also needed so that routine changes can be made to the protection strategy to ensure that it remains effective. Fundamental to risk management is identification of residual risk with a network protection strategy in place and acceptance of that risk by management.

15. Establish a network protection strategy based on the principle of defense in depth.
    A fundamental principle that must be part of any network protection strategy is defense in depth. Defense in depth must be considered early in the design phase of the development process and must be an integral consideration in all technical decision-making associated with the network. Utilize technical and administrative controls to mitigate threats from identified risks to as great a degree as possible at all levels of the network. Single points of failure must be avoided, and cybersecurity defense must be layered to limit and contain the impact of any security incidents. Additionally, each layer must be protected against other systems at the same layer. For example, to protect against an inside threat, restrict users to access only those resources necessary to perform their job functions.

16. Clearly identify cybersecurity requirements.
    Organizations and companies need structured security programs with mandated requirements to establish expectations and allow personnel to be held accountable. Formalized policies and procedures are typically used to establish and institutionalize a cybersecurity program. A formal program is essential for establishing a consistent, standards-based approach to cybersecurity throughout an organization and eliminates sole dependence on individual initiative. Policies and procedures also inform employees of their specific cybersecurity responsibilities and the consequences of failing to meet those responsibilities. They also provide guidance regarding actions to be taken during a cybersecurity incident and promote efficient and effective actions during a time of crisis. As part of identifying cybersecurity requirements, include user agreements and notification and warning banners. Establish requirements to minimize the threat from malicious insiders, including the need for conducting background checks and limiting network privileges to those absolutely necessary.
17. Establish effective configuration management processes.
    A fundamental management process needed to maintain a secure network is configuration management. Configuration management needs to cover both hardware configurations and software configurations. Changes to hardware or software can easily introduce vulnerabilities that undermine network security. Processes are required to evaluate and control any change to ensure that the network remains secure. Configuration management begins with well-tested and documented security baselines for your various systems.
18. Conduct routine self-assessments.
    Robust performance evaluation processes are needed to provide organizations with feedback on the effectiveness of cybersecurity policy and technical implementation. A sign of a mature organization is one that is able to self-identify issues, conduct root cause analyses, and implement effective corrective actions that address individual and systemic problems. Self-assessment processes that are normally part of an effective cybersecurity program include routine scanning for vulnerabilities, automated auditing of the network, and self-assessments of organizational and individual performance.
19. Establish system backups and disaster recovery plans.
    Establish a disaster recovery plan that allows for rapid recovery from any emergency (including a cyberattack). System backups are an essential part of any plan and allow rapid reconstruction of the network. Routinely exercise disaster recovery plans to ensure that they work and that

personnel are familiar with them. Make appropriate changes to disaster recovery plans based on lessons learned from exercises.

20. Senior organizational leadership should establish expectations for cybersecurity performance and hold individuals accountable for their performance.
    Effective cybersecurity performance requires commitment and leadership from senior managers in the organization. It is essential that senior management establish an expectation for strong cybersecurity and communicate this to their subordinate managers throughout the organization. It is also essential that senior organizational leadership establish a structure for implementation of a cybersecurity program. This structure will promote consistent implementation and the ability to sustain a strong cybersecurity program. It is then important for individuals to be held accountable for their performance as it relates to cybersecurity. This includes managers, system administrators, technicians, and users/operators.

21. Establish policies and conduct training to minimize the likelihood that organizational personnel will inadvertently disclose sensitive information regarding SCADA system design, operations, or security controls.
    Release data related to the SCADA network only on a strict need-to-know basis, and only to persons explicitly authorized to receive such information. "Social engineering," the gathering of information about a computer or computer network via questions to naive users, is often the first step in a malicious attack on computer networks. The more information revealed about a computer or computer network, the more vulnerable the computer/network is. Never divulge data revealed to a SCADA network, including the names and contact information about the system operators/administrators, computer operating systems, and/or physical and logical locations of computers and network systems over telephones or to personnel unless they are explicitly authorized to receive such information. Any requests for information by unknown persons need to be sent to a central network security location for verification and fulfillment. People can be a weak link in an otherwise secure network. Conduct training and information awareness campaigns to ensure that personnel remain diligent in guarding sensitive network information, particularly their passwords.

REFERENCES AND RECOMMENDED READING

Associated Press (AP). (2009, April 4). "Goal: Disrupt." *Virginian-Pilot* (Norfolk).

Brown, A. S. (2008). "SCADA vs. the Hackers." American Society of Mechanical Engineers.

DOE. (2001). *21 Steps to Improve Cyber Security of SCADA Networks*. Washington, DC: Department of Energy.

Ezell, B. C. (1998). *Risks of Cyber Attack to Supervisory Control and Data Acquisition*. Charlottesville: University of Virginia.

FEMA. (2008). *FEMA 452: Risk Assessment: A How-To Guide*. https://www.wbdg.org/FFC/DHS/ARCHIVES/fema452.pdf, accessed May 17, 2023.

FEMA. (2015). *Protecting Critical Infrastructure against Insider Threats*. https://emilms.fema.gov/is_0915/curriculum/1.html, accessed May 17, 2023.

FBI. (2000). *Threat to Critical Infrastructure*. Washington, DC: Federal Bureau of Investigation.

FBI. (2007). *Ninth Annual Computer Crime and Security Survey*. Washington, DC: Computer Crime Institute and Federal Bureau of Investigations.

FBI. (2014). *Protecting Critical Infrastructure and the Importance of Partnerships*. https://www.fbi.gov/news/speeches/protecting-critical-infrastructure-and-the-importance-of-partnerships, accessed May 17, 2023.

GAO. (2003). *Critical Infrastructure Protection: Challenges in Securing Control System*. Washington, DC: U.S. Government Accountability Office.

GAO. (2015). *Critical Infrastructure Protection: Sector-Specific Agencies Need to Better Measure Cybersecurity Progress*. Washington, DC: U.S. Government Accountability Office.

Minter, J. G. (1996). "Preventing Chemical Accidents Still a Challenge." *Occupational Hazards*, September.

National Infrastructure Advisory Council. (2008). *First Report and Recommendations on the Insider Threat to Critical Infrastructure*. Washington, DC.

NIPC. (2002). *National Infrastructure Protection Center Report*. Washington, DC: National Infrastructure Protection Center.

Spellman, F. R. (1997). *A Guide to Compliance for Process Safety Management/Risk Management Planning (PSM/RMP)*. Lancaster, PA: Technomic Publishing.

Stamp, J., et al. (2003). *Common Vulnerabilities in Critical Infrastructure Control Systems*. 2nd ed. Sandia National Laboratories.

USDOE. (2002). *Vulnerability Assessment Methodology: Electric Power Infrastructure*. Washington, DC: U.S. Department of Energy.

USDOE. (2010). *Energy Sector-Specific Plan: An Annex to the National Infrastructure Protection Plan*. Washington, DC: U.S. Department of Energy.

USEPA. (2005). "EPA Needs to Determine What Barriers Prevent Water Systems from Securing Known SCADA Vulnerabilities." In J. Harris, Final Briefing Report. Washington, DC: U.S. Environmental Protection Agency.

Warwalking. (2003). http://warwalking.tribe.net, accessed May 9, 2018.

Wiles, J., et al. (2007). *Techno Security's Guide to Securing SCADA*. Burlington, MA: Elsevier.

Young, M. A. (2004). *SCADA Systems Security*. SANS Institute.

Chapter 6

Cybersecurity Risks and Scenarios

EMERGENCY SERVICES SECTOR CYBER RISK PROFILE

Scenarios provide the primary means to identify and rate risks. Risk scenarios represent a distinct set of events that could significantly affect the ESS's ability to effectively perform its responsibilities. The scenarios allow participants to discuss practical real-life situations and determine threats, vulnerabilities, and consequences while allowing for consideration of compounding and overlapping effects often spreading across the multiple ESS disciplines' cyber infrastructures.

Scenario 1: Natural Disaster Causes Loss of 911 Capabilities

Natural disasters are threats to ESS disciplines and their cyber infrastructure. Natural disasters typically affect specific geographic locations or regions and cause immediate impacts or degradation in normal day-to-day ESS cyber infrastructure and communications capabilities, including 911 capabilities. This scenario would have compounding consequences. Any natural disaster significant enough to render 911 communications inoperable is also likely to cause damage to property and potentially injury or loss of life to persons in the surrounding communities. The most catastrophic dimension of this scenario is the case where numerous customers are in need of 911 for assistance and the capability is unavailable via traditional communications means such as telephone. In this case, customers would be required to identify alternate methods of communication with ESS entities, or they would be required to physically go to the ESS entities' facilities.

Specific disasters were not selected and rated because of the large geographic footprint of the United States; instead, the ESS-CRA evaluation assumes that a disaster that is most likely to affect a particular region or locale is significant enough to be deemed a federally declared natural disaster.[1]

The components of this scenario include undesired consequences, the vulnerabilities that can lead to those undesired consequences, and the threats that can exploit those vulnerabilities. The components are relevant to the following ESS disciplines: law enforcement, fire and emergency services, EMS, public works, and public safety communications and coordination (PSC&C). The following analysis evaluates the risks and identifies the impacts of a natural disaster causing the loss of 911 capabilities to the ESS disciplines and its cascading impacts.

When a telecommunications system, such as 911 systems or CAD systems, are degraded or inoperative, situational awareness by all parties across all disciplines is significantly affected, thereby reducing the likelihood of successful coordinated responses to emergencies.

The PSC&C, EMS, fire and emergency services, and law-enforcement disciplines are more negatively affected than the other two ESS disciplines—public works and emergency management. If a natural disaster, including meteorological, geological, or biological incidents, disables one or several PSAPs, or otherwise causes the loss of 911 capabilities, the operational capabilities across all ESS disciplines—including law enforcement, fire and emergency services, PSC&C, public works, and EMS—are put at risk. The consequences of the loss or degradation of 911 capabilities can cascade across several different critical infrastructure sectors and significantly affect the ability of ESS to perform emergency response.

It is a common practice across the ESS community to incorporate the concept of redundancy in 911 infrastructures. However, redundancy is not helpful in the face of significant natural threats if the redundant infrastructures are located in the same geographic area, as is the case for the majority of PSAPs. Backup PSAP functions may be limited to one PSAP located close to the primary PSAP. This scenario considers the possibility that natural disasters may be destructive enough to render both the primary and the redundant PSAPs insufficient to provide emergency communications services.

ESS entities provide essential services that seek to limit loss of life and damage. The degradation of ESS disciplines results in exposure of ESS entities' consumers to increased risks, leaving them to mitigate or respond to the situation themselves. This situation can trigger cascading consequences to public health and safety as well as greater economic and physical destruction.

As previously noted, natural disasters are some of the most regularly occurring threats to the ESS in general, and to the PSC&C discipline specifically, because 911 capabilities are among the discipline's core services. The capabilities of 911 depend heavily on commercial communications infrastructure.

Many citizens use plain old telephone service (POTS) and wireless networks to connect to 911, and many public safety call centers use these networks to connect to citizens, send critical dispatches, and coordinate efforts with first responders. Natural disasters frequently disrupt, damage, or destroy infrastructure in the commercial networks needed by the PSC&C discipline. In these scenarios, citizens may lose the ability to make calls (and therefore reach 911), call centers may have outages caused by damage to their communications infrastructure or lines, or communications lines may become overwhelmed and lead to denial of service.

The natural threats that could cause this scenario are numerous and vary by region of the country, but some of the more prominent threats include earthquakes, hurricanes, and winter storms. Other threats may include celestial events, such as geomagnetic or solar radiation storms that cause communications outages, or biological threats, such as epidemics or pandemics, which might not cause outages but could still degrade conditions as a result of resources being overwhelmed. Several key vulnerabilities could be exploited by these threats. Physical vulnerabilities include the concentration of assets and/or cyber infrastructure (e.g., telecommunications hotel or call centers), the location of key facilities and assets (e.g., communications lines, PSAPs, cell towers) in vulnerable geographic locations, and a lack of alternate routing for 911 call centers. There are also technological vulnerabilities, such as poorly designed architectures with single points of failure or a reliance on commercial providers, and process-related vulnerabilities, such as a lack of preparation, training, or exercises. The assessment of this scenario rated each of these threats and vulnerabilities as high, resulting in a high likelihood of failure. In fact, this scenario had the highest likelihood of failure in the PSC&C discipline because these issues arise more frequently than in any of the other scenarios.

The direct consequence to the PSC&C is the loss of the ability to coordinate emergency response efforts or to deploy resources effectively. In addition, the consequences may cascade and be felt by other ESS disciplines and other sectors beyond the ESS. Most notably, PSC&C-sector service providers in the affected area would be required to deploy resources to restore lines and infrastructure that are critical to 911 services. The health-care and public health (HPH) sector would also be affected, as it would be more difficult to provide emergency medical services.

The impact of this scenario is likely to be felt in a regional or local area as opposed to at the national level. However, because of the severe impact and cascading effects that can occur in those affected local regions, the consequences were assessed to be high. Since this scenario has a high likelihood of occurrence and high consequences, this risk was rated higher than any other risk.

NATURAL THREAT: FIRE AND EMERGENCY SERVICES

Natural disasters create hardship and heartache for every person living or working in the affected area. Those persons experiencing the effects of a natural disaster firsthand may be facing the very worst moment of their lives as they find themselves or others in need of immediate assistance, whether it is the result of severe weather, earthquake, tidal wave, wildfire, or another naturally occurring phenomenon. Floods can submerge sensitive electronic equipment at telephone infrastructure sites. Wildfires can burn overhead telephone lines or spread from wooded areas into central office facilities. Earthquakes can sever buried telephone lines, topple telephone poles, or snap lines stressed by excessive movement. While these are some of the more prominent natural disasters to which telephone systems are vulnerable, they are not the only ones. If the infrastructure facilities have not been hardened against the effects of a natural disaster, or if the infrastructure itself lies in a geographical location that is susceptible to natural disasters (such as floods or earthquakes), with insufficient telephone-route diversity and alternate switching, the carriers responsible may be unable to quickly restore or reconstitute the level of service required to serve a given populace.

People who find themselves in need of help from one or more of the ESS disciplines require the ability to contact those disciplines. While some urban areas may retain fireboxes on their downtown street corners, and some buildings may provide manual pull stations designed to alert building occupants to evacuate while they also transmit an emergency signal to a fire alarm communications center, most people require a dial tone and telephone connectivity to summon their nearest fire and emergency services organization. For those people, their ability to notify authorities of their needs is tied to their local emergency number, or 911.

Connectivity and a dial tone are resources provided by a competitive local exchange carrier to dial and connect with 911. To reach a 911 call taker, a caller must be able to use a landline or mobile telephone or an enhanced special mobile radio device to dial the emergency number. If the caller has no dial tone, the call cannot be connected. If the infrastructure that supports the dial tone is damaged or destroyed, the call can neither be connected nor completed because components, such as telephone lines, central offices, or switches, are essential to carrying the call to its destination. As a result, callers may be left panicked and confused about what to do. They may not have the knowledge of the location of—or the ability to send for help to—the nearest fire and emergency services station to summon firefighters or emergency response personnel in person.

In addition to the communications infrastructure vulnerabilities that may be revealed in the wake of a natural disaster, the fire and emergency services discipline has a critical dependency on the communications sector; in the event of a natural disaster, reconstitution of communications infrastructures is performed by the communications sector. However, there are similar risks that fire and emergency services organizations can experience. If the loss of 911 capabilities is caused by damage to or the destruction of the 911 communications center, provisions may have been made to develop alternate locations to which 911 calls can be routed. However, for the majority of PSAPs, redundancy is extremely limited geographically, and the location of an alternate PSAP may be vulnerable to the same natural disaster that affected the telephone companies. Some organizations, particularly those that have migrated to digital, Internet Protocol (IP)–based infrastructure, may already have alternate communications centers in place, but if they have not properly planned for the activation and maintenance of such facilities, they may find themselves ill prepared to reestablish 911 services. Further, if the personnel expected to use alternate facilities have not been trained to migrate services to the alternate site and have never conducted exercises to develop proficiency in doing so, the chances of successfully transitioning from the damaged site to the alternate site are also significantly reduced.

Fire and emergency services organizations in parts of the country where natural disasters occur frequently may experience a higher likelihood of losing 911 capabilities. In California, for example, wildfires, earthquakes, high winds, and tsunamis present ongoing threats; however, there are no parts of the country in which a natural disaster has not, and will not again, occur. If a citizen is unable to reach the local fire and emergency services organization when needed, and the organization is unable to provide for connectivity with the communities they serve, the consequences place lives and property that have already been threatened by the natural disaster itself at an even greater level of risk. It is this high likelihood of a natural disaster's occurrence causing a loss of 911 capabilities that, when coupled with the severe consequences, make this the greatest threat to the fire and emergency services discipline.

Scenario 2: Lack of Availability of Sector Database Causes Disruption of Mission Capability

ESS cyber infrastructure includes databases and their supporting elements. ESS databases are critical to supporting sector missions and activities. Should a database be unavailable, there will be disruption to mission capabilities within and across ESS disciplines. Databases are vulnerable to cyberattack and subject to man-made deliberate and man-made unintentional threats.[2]

The components of this scenario include undesired consequences, the vulnerabilities that can lead to those undesired consequences, and the threats that can exploit those vulnerabilities. The components are relevant to all ESS disciplines but focus on the law-enforcement and fire and emergency services disciplines. Notably, in this scenario, law-enforcement personnel could lose the ability to run accurate criminal background checks on suspects, jeopardizing both their own safety as well as that of the general public. Law-enforcement agencies would have to rely on more traditional methods of identification and criminal background checks such as using radio communications or mobile communications and relying on manual checks of state and local records. In addition, firefighters and emergency service personnel could lose access to data that would assist them in responding to an emergency, such as floor-plan layouts, commercial building contents, documented hazards, and protocols and best practices to safely deal with various hazards. Firefighters and emergency service personnel would have to rely on printed materials on file, which are often out of date or inaccessible.

Risk Assessment Scenario 2: Lack of Availability of Sector Database Causes Disruption of Mission Capability

A man-made deliberate attack against ESS databases is most likely to occur within the PSC&C, law-enforcement, emergency management, and public works disciplines. The PSC&C discipline would be most affected by such an incident, along with the public works discipline, which has a higher consequence rating than in the man-made deliberate scenario. The law-enforcement discipline would experience relatively equal effects on its mission. In addition, in the man-made unintentional threats scenario, the EMS and fire and emergency services disciplines are at risk. The emergency management discipline's likelihood and consequence ratings are lower in the unintentional threat scenario.

MAN-MADE DELIBERATE THREAT: PUBLIC SAFETY COMMUNICATIONS AND COORDINATION/FUSION (PSC&C)

The PSC&C discipline relies on various kinds of ESS databases, including criminal justice databases, such as the National Crime Information Center (NCIC) database; geospatial databases with critical infrastructure data; and motor vehicle administration databases. These databases are most notable in the subdisciplines of providing fusion center capabilities and providing GIS and CAD capabilities. While these databases are often owned and maintained by other ESS disciplines, such as public works, law enforcement, and fire and emergency services, the PSC&C discipline uses them to coordinate effective response operations among all of the ESS disciplines.

Potential threat actors who could deliberately cause this scenario include criminals, activist hackers (hacktivists), cyber vandals, and/or corrupt or disgruntled insiders. Malicious actors can be motivated by objectives that include obstruction, counterintelligence, and deception, such as trying to embarrass an agency, looking for thrills, diverting attention from or adding to the magnitude of a separate attack, or eliminating certain records. Such actors may be part of a structured organization and have operational knowledge of technology that can be used to attack databases.

There are database vulnerabilities that malicious actors try to identify and take advantage of, especially if the database is linked to web-based applications. According to the SANS (SysAdmin, Audit, Network, Security) Institute, attacks against web applications constitute more than 60 percent of all attack attempts observed on the internet, and SQL injections against databases account for nearly one-fifth of all security breaches.[3] In addition, several available open-source database vulnerability tools make it easy for the malicious actor to probe systems. The actor does have constraints, such as the need to be covert, a small window of time to execute the attack, and often a lack of insider physical or logical access to systems that include law-enforcement databases. However, the capabilities and resources of this threat actor, combined with database vulnerabilities, make the malicious threat actor a key concern of the law-enforcement discipline—mostly because of the consequences that can result from successful degradation of law-enforcement personnel's access to the databases.

As a result, the risk likelihood associated with these scenarios ranges from low to medium, and the consequences range from low to medium-high. The man-made deliberate threat scenarios are higher impact than the man-made unintentional threat scenarios. The likelihood of an attack can also rise if a man-made deliberate threat actor targets a database with vulnerabilities caused by man-made unintentional threats such as untrained users.

While the relative consequence of this scenario can vary depending on the specific database targeted and the magnitude of the attack, the consequences for the PSC&C discipline could be very significant. Fusion centers and call centers are expected to maintain a high level of performance at all times, and while these centers often have offline or paper backups, using those resources may slow or degrade their ability to deploy or coordinate efforts. For example, the loss of a GIS database may prevent a PSAP from identifying the location of a 911 caller, thereby slowing response time until a location can be identified manually. The loss of key databases may also prevent a dispatcher from receiving or responding to requests that are pending action in a CAD system. Consequences may also cascade to other critical infrastructure sectors, especially industries with a more critical dependence on response capabilities because of the nature of their activities (such as certain types of manufacturing or work with HAZMAT).

Given the serious consequences in this scenario, the impact could be significant, especially if it occurred in a highly populated area. The impact will vary based on the length and severity of the outage and could be mitigated through maintaining proper backups and effective training on how to respond during such an outage. In addition, it is important to note that while the risks from man-made deliberate and unintentional threats are similar, the subject-matter experts who provided input to this assessment judged that the likelihood of an unintentional threat causing the outage was slightly greater because of existing mitigations such as access control mechanisms and the greater access that authorized insiders have to such databases.

MAN-MADE UNINTENTIONAL THREAT: EMERGENCY MEDICAL SERVICES

The greatest threat to ESS databases are unintentional acts, such as software design defects or programming failures. Other threats include the incorrect input of data or inaccurate modification or deletion of records. The latter circumstance can affect other mission-critical services, such as properly recording and saving patient-care data or providing accurate medical billing. Another source for compromise can be unintentional disruption of critical databases. For example, the National EMS Information System (NEMSIS) is a national effort to promote the development of local, state, and national EMS patient electronic health-care records and data systems.[4] At the national level, the goal of NEMSIS is to maintain a national EMS database. As of February 2012, the majority of states and territories have implemented a NEMSIS-based state EMS data system, with thirty-six states and territories submitting NEMSIS data to the national EMS database.

As described below, people and process vulnerabilities are predominantly unintentional but can lead to compromised EMS databases.

- EMS providers, other employees, or third-party software and database specialists are the most likely persons to accidentally cause a database compromise. They may do so via acts of carelessness or recklessness, but other factors such as insufficient training or even power outages during programming or data entry can place these persons in the position of accidentally causing a database compromise. For example, database users can inadvertently corrupt their information when they reboot or shut down their access terminals while the database is open.

- Physical and environmental security factors, such as where and how access terminals are placed or used, can contribute to unintentional database compromises. Poor physical access control, exposure of equipment or software to trip hazards or drops, and lack of protection of power equipment are examples of physical and environmental security factors that can lead to disruption of information systems, including databases.

Depending on which database is compromised, the consequences will vary. It may take a short time to recognize a compromise in a database function, but it may take hours to reconstitute the data stored therein, unless effective redundancy is available. Even then, the cascading effects of some compromises, such as to medical billing, can delay vital revenue streams that sustain some EMS agencies. Other compromises may have more critical implications. For example, if the Master Street Address Guide in a CAD system is compromised, EMS response may be delayed, or the request may be directed to the wrong agency or jurisdiction. If geospatial systems are compromised, such as automatic vehicle location, it is possible that the closest EMS resources to an emergency may be overlooked and a more distant unit sent in its place. If online resources such as poison control experience a database compromise, the consultation may be inaccurate and the treatment recommended either unreliable or even contraindicated.

Given these possibilities, while the risk of compromising a critical EMS database is low, the consequences for an EMS agency, and perhaps even for its patients, are high.

Scenario 3: Compromised Sector Database Causes Corruption or Loss of Confidentiality of Critical Information

As stated in the previous scenario, ESS databases are critical to supporting sector missions and activities. In the case of a compromised sector database causing corruption or loss of confidentiality of critical information, there will be disruption to mission capabilities.

Corruption or loss of confidentiality of critical information residing on a database can be the result of a man-made deliberate threat (e.g., hacktivist, cybercriminal) or a man-made unintentional threat (e.g., software or programming failure). Databases are frequently targeted in man-made intentional attacks. If those databases are not protected against unauthorized access and modification and are not regularly backed up, data loss or corruption can significantly impair ESS mission capabilities. The threats, vulnerabilities, and consequences in this scenario affect all six critical ESS disciplines. The ESS baseline risk assessment report assesses the impacts of the scenario on six ESS disciplines and provides the foundation for creating risk mitigation strategies.

Intentional or unintentional manipulation of law-enforcement databases that causes or leads to data loss or corruption can jeopardize the operations of law-enforcement resources such as CAD and criminal justice networks and systems. Detecting such an incident that affects availability will likely take a matter of minutes in the case of data corruption, but it may take hours to return service to a minimally acceptable level. An incident that corrupts the information of a database may take longer to recover from, depending on the frequency and availability of information backups. Loss of data confidentiality may take longer to detect depending on the sophistication and goals of the threat actor. If the threat is unintentional, the impact may be mitigated quickly by undoing actions that caused the incident or by restoring data.

An incident occurring as a result of a deliberate threat actor may take longer to mitigate. Depending on the databases and associated systems attacked, the impacts of an attack may cascade to impact areas such as homeland security. For example, the court system may be affected by an attack on related databases, foreign operations with Interpol may be delayed, and Medicaid billing/financial management can be impaired. Public health and safety may be affected if law-enforcement and fire databases impair fire and search-and-rescue missions.

As noted in the analysis for scenario 2, a malicious actor can use numerous database vulnerabilities, especially if the database is linked to web-based applications. The man-in-the-middle attack is a common hacker attack and can be used to acquire database credentials. There are also several available open-source database vulnerability tools that make it easy for the malicious actor to probe systems.

The components of scenario 3 include undesired consequences, the vulnerabilities that can lead to those undesired consequences, and the threats that can exploit those vulnerabilities. The components are relevant to the following ESS disciplines: law enforcement, fire and emergency services, EMS, emergency management, public works, and PSC&C. A compromised sector database that causes the corruption or loss of confidentiality of critical information can have a significant impact across the ESS. Incorrect or corrupt data can place public safety agencies, first responders, and the general public in significant jeopardy.

Law-enforcement agencies, in particular, rely extensively on the accuracy of criminal and other public safety databases used to verify a person's identity and/or criminal record. If this information were to become corrupted or be stolen, law-enforcement processes could be significantly delayed, and criminal organizations may be enabled to perform counterintelligence operations. Moreover, if frontline first responders feel that the information they are getting from various public safety databases is in some way inaccurate, they will lose confidence in the system, thereby rendering it less effective.

The other ESS disciplines are affected by the scenario because—similar to law enforcement—they rely on shared data sources and information to improve their situational awareness, coordinate response and operations, and manage communications and coordination tools. For example, each discipline uses GIS systems as a tool in conducting their activities. If GIS data is inaccurate or corrupted—deliberately or unintentionally—response agencies could be dispatched to the wrong destination, thereby putting the entity or individual needing the emergency response at risk of further harm.

The public works discipline uses databases to support capabilities such as geospatial tools and systems that store critical infrastructure information. These types of databases store information related to utility placement, facility types and locations, vehicle locations, and navigational aids. Other databases may store information related to resource inventories, employee records, or agency payrolls. The information contained in these databases can be analyzed to identify resource needs or infrastructure disruption trends to ensure that function is properly maintained. The increased connectivity provided by IP-based systems and wireless devices has now made it possible for this information to be shared with connected devices, such as computers, smartphones, and GPS units, to provide public works employees with instant access to the data needed to sustain the function's operations. In addition, these devices are able to record and transmit information to the databases, ensuring that the information contained within is updated and accurate.

Although the emergency management discipline can support some public works discipline operations, such as field operations, during an emergency, the corruption of critical information resulting from a database compromise could cause redirection of public works resources and slow incident response times. Dispatches could be misdirected, or utility services, such as electricity, water, or wastewater, could be lost depending on the type of database involved. The increased adoption of IT capabilities, and the corresponding IT security culture, has increased the discipline's awareness of

IT-related incidents so that detection of such an event could be measured in hours. The rapid detection of a compromised database allows for recovery within a few hours, and reconstitution of the discipline's operations can be achieved within a few days. The cascading impacts on other sectors would be slight and limited to those sectors that play a role in utility services, such as the energy and water sectors. The database vendor and the agency involved in the incident would face substantial public confidence impacts in the local area affected but not at the national level.

Risk Assessment Scenario 3: Disciplines and Cyber Infrastructure Affected

In the case of a man-made deliberate scenario, the likelihood of the threat exploiting the vulnerability is generally low, but the consequence varies from low to high across the ESS disciplines. The PSC&C and law-enforcement disciplines have a higher consequence because deliberate corruption or stealing of data may have a more significant impact on day-to-day operations. For example, corrupted data in a law-enforcement database or information leaked to the public can delay or impair criminal investigations. In the case of a man-made unintentional threat, the likelihood of the scenario is lower for the law-enforcement and emergency management disciplines because of the emphasis on integrity of data.

Conversely, the likelihood of a man-made unintentional threat is higher for the fire and emergency services, EMS, public works, and PSC&C disciplines because of their transactional time-sensitive data operations. Overall, the consequences for man-made unintentional threats are slightly lower because databases have built-in processes to prevent unintentional deletion or modification of large amounts of data, and data is generally updated periodically.

MAN-MADE DELIBERATE THREAT: LAW ENFORCEMENT

The ESS law-enforcement discipline uses databases to assist in several activities, including managing personnel and equipment, conducting criminal investigations, gathering and protecting evidence, and apprehending perpetrators of crimes. Databases support cyber-technology resources like CAD and criminal justice networks and systems. For example, in criminal justice networks and systems, databases support analysis to find commonalities that may have investigative value, such as patterns in sites of crimes or names that recur from contacts at crime scenes. Fingerprint and identification databases provide rapid analysis of electronic print cards to search for suspects with prior criminal justice contacts. Databases are used by law-enforcement agencies to enter, modify, and withdraw data. The NCIC, for example, requires that the state police agency in every state audit the local, state, and tribal law-enforcement agencies to assure that only staff that is trained, certified, and authorized are accessing the system, that recordkeeping is organized and up to date, and that all entries meet FBI criteria.

Intentional or unintentional disruption of law-enforcement databases can jeopardize the operations of law-enforcement resources such as CAD and criminal justice networks and systems. The time to detect such an incident that affects availability will likely be a matter of minutes, but it may take hours to return service to a minimal acceptable level. If the threat source is unintentional, the impact may be mitigated quickly by undoing actions that caused the incident or by restoring data. An incident that occurs as a result of a deliberate threat actor may take longer to mitigate because of the threat actor's/actors' desire to conduct the attack undetected.

Attacks on law-enforcement databases may be used as a basis for attacks on more secure networks such as Law Enforcement Online. Depending on the databases and associated systems attacked, the impacts of an attack may cascade to other critical infrastructure sectors. For example, the court system may be affected by an attack on related databases, foreign operations with Interpol may be delayed, and Medicaid billing/financial management may be impaired. Public health and safety may be affected if law-enforcement databases impair fire or search-and-rescue missions. If the attack is reported publicly, it may reduce confidence in the government to perform its essential tasks. In addition, if the attack is reported publicly, it may reduce confidence in the government to regulate law enforcement.

This analysis focuses on deliberate threat actors because the consequences of a deliberate threat actor will likely result in a higher consequence to the sector. Major potential man-made deliberate threats to databases shared by multiple jurisdictions are cybercriminals and organized crime. These malicious actors would be motivated by objectives that include obstruction, counterintelligence, and deception. Such actors would likely be part of a structured organization and have operational knowledge of technology that could be used to attack databases. The actors do have constraints, such as the need to be covert, a small window of time to execute the attack, and often a lack of insider physical or logical access to systems, including law-enforcement databases. However, the capabilities and resources of these threat actors, combined with database vulnerabilities, make this a significant concern of the law-enforcement discipline.

Like cybercriminals, organized crime actors can be motivated by objectives that include obstruction, counterintelligence, and deception. Such actors may be part of a structured organization and have operational knowledge of technology that can be used to attack databases. Cyber vandals or radical activists are the additional threat actors associated with man-made deliberate incidents described in this scenario. Thieves may also be interested in the proprietary information contained within some databases, such as those within public works agencies. For cyber vandals and radical activists, or hacktivists, corrupting a public works database could be a means to exert influence over government operation and draw attention to their cause (e.g., hacking groups LulzSec and Anonymous).

Such threat actors typically understand the underlying technology, tools, and methods needed to compromise and corrupt a database and can quickly create new attacks to adapt to the database involved. As discussed in scenario 2, there are several database vulnerabilities that malicious actors can identify and leverage. These vulnerabilities not only lead to loss of availability but can also result in compromise of database integrity and confidentiality. For example, the hacker group LulzSec claimed that it used an SQL injection attack against Sony Pictures in 2011 to steal user data of more than thirty-five thousand users.

The risk likelihood associated with these scenarios ranges from low to medium, and the consequences range from low to medium-high. The man-made deliberate threat scenarios are higher impact than the man-made unintentional threat scenarios. The relative consequences can vary depending on the specific database targeted and the magnitude of the attack. The likelihood of an attack can also increase if a man-made deliberate threat actor targets a database with vulnerabilities caused by man-made unintentional threats such as untrained users.

MAN-MADE UNINTENTIONAL THREAT: EMERGENCY MEDICAL SERVICES

Unlike the other disciplines in the ESS, EMS agencies operate under a variety of models that include for-profit corporations, nonprofit corporations, and not-for-profit organizations. The need to capture critical data to support these models is just as varied and depends on whether the agency charges patients (directly or indirectly) for their services. The databases that agencies using these models build, maintain, and operate may support

- Business interests such as accounts receivable and accounts payable and fleet-maintenance records
- Patient care such as electronic patient-care reporting systems and approved medical treatment protocols

- Defined service areas, such as a Master Street Address Guide
- Receiving facilities such as status of capabilities and points of contact for freestanding emergency facilities, general hospitals, and specialized sites such as burn centers
- Training and licensing such as qualifications for personnel, certifications and expiration dates, and licenses to practice

An example of an EMS database is NEMSIS.[5] As mentioned in scenario 2, the goal of NEMSIS is to develop and maintain a national EMS database that includes local, state, and national EMS patient electronic health-care records and data systems.

Although the risks that an EMS database may face are similar to the risks that other functional databases may encounter, the emphasis of many of the databases on which EMS agencies rely is limiting liability to the agency and promoting an effective and accurate accounting of patient care when transferring responsibility for a patient to another EMS agency. One scenario in which this could be important is when an ambulance crew delivers a critically injured patient to a landing zone so that a waiting helicopter can provide rapid medical evacuation to a trauma center. This would also be important at a receiving facility when patient care is turned over to a physician or nurse in a hospital's emergency department. While these interests may be behind the creation of many EMS databases, the business interests that a database supports may include the recordkeeping necessary to remain solvent through accurate billing and accounting for payments received. Thus, when any of these databases is corrupted, the loss of access to essential data can affect the ability of an agency to pay for fuel, supplies, and insurance so that its ambulances can respond to calls; the ability to identify the best hospital to which a patient in need of specialized care should be transported; and the ability to optimize the pre-hospital care that trained and equipped emergency medical care providers can deliver en route to a hospital. Finally, if the patient-care reporting system is corrupted, it can delay or cause deferment of better care for very ill or seriously injured patients because hospital staff cannot reliably account for the medical interventions taken by emergency medical care providers before their arrival with the patient.

There are several likely sources of an unintentional compromise of a database that causes the corruption of critical information. The sources of these vulnerabilities may include

- Personnel: EMS providers, agency administrators, and third-party contractors (e.g., billing agencies, database and software engineers and administrators, and hardware technicians) may inadvertently cause database corruption when they use incorrect keystrokes, make inaccurate entries or

modifications of data, or use a corrupted formula that misaligns or otherwise renders data unreliable.
- Processes: Improperly trained personnel, personnel who are fatigued from emotionally and physically demanding incidents, personnel who may be working under highly stressful conditions, and personnel who are operating under poorly maintained or enforced security policies have been exposed to a process vulnerability that could cause an unintentional database corruption.

The threat of a database being unintentionally corrupted stems from sources such as poor training or trained personnel applying their skills improperly, carelessly, or even recklessly. A technical flaw may also unintentionally corrupt a database. Other sources of corruption include power failures or surges, and even lightning strikes.

Fortunately, the likelihood of an unintentional corruption to a database containing critical information is low; however, the consequences of such instances are very high. Agencies could lose revenues essential to operating a viable EMS delivery system; but more significantly, the affected EMS agency could be exposed to civil liability, and the public or the hospitals they serve could be exposed to errors in patient care.

Scenario 4: Public Alerting and Warning System Disseminates Inaccurate Information

Public alerting and warning systems contribute to several ESS disciplines' operational capabilities. These systems range from the national-level Integrated Public Alert Warning System for major emergencies to regional and local alert and warning systems. These systems provide alerts for a variety of events and ESS disciplines. Both older and newly modernized public alerting and warning systems can be at risk for either intentional or unintentional dissemination of inaccurate information. The ESS disciplines targeted in this scenario are emergency management, PSC&C, public works, and law enforcement.

The components of this scenario include undesired consequences, the vulnerabilities that can lead to those undesired consequences, and the threats that can exploit those vulnerabilities. When a public alerting and warning system disseminates inaccurate information, the "share, communicate, notify" portion of the value chain is the most severely affected. When a public alerting and warning system disseminates inaccurate information, it not only creates unnecessary panic, but it can also cause the public to lose confidence in the various public alerting systems and lead to the public ignoring an actual emergency because it is believed to be false. If this occurs, emergency managers will need to use other less efficient, improvised methods such as door-to-door notification, live TV, and/or radio announcements.

Risk Assessment Scenario 4: Public Alerting and Warning System Disseminates Inaccurate Information

In the case of a man-made deliberate threat in this scenario, the likelihood of the threats affecting the ESS disciplines is low to medium, with an emphasis on the law-enforcement discipline and emergency management, which in this scenario will likely have to deal directly with the event and subsequent reactions. The relative consequences for several of the disciplines, including law enforcement, emergency management, and PSC&C, is high because the public is likely to believe and react to information from a public alerting and warning system.

In the case of a man-made unintentional threat, the likelihood of the risks in the scenario is generally higher for the ESS disciplines, and the consequences are roughly the same or slightly lower. A man-made unintentional threat may be more common because of the potential for human error and technical error. However, unintentional threats may be detected and mitigated more quickly if the user associated with the unintentional threat is immediately aware of the incident.

MAN-MADE DELIBERATE THREAT: EMERGENCY MANAGEMENT

Emergency management professionals rely on a small number of systems to perform their disciplines, and public alert and warning systems are one of the more important types. The emergency management community is the "power user" group of public alerting and warning systems. They use systems such as an emergency alert system (EAS), emergency alert networks that disperse information to citizen subscribers via their registered devices (e.g., cellular telephones and pagers), sirens, and public address systems. Some agencies use reverse 911 or similar products to send warnings to entire communities when the need arises. Public alerting and warning systems primarily support the following core activities of the emergency management value chain: share, communicate, notify; share, respond, operate; and record, save, review.

These alerting and warning systems use software that may be dependent on other cyber communications resources to disseminate their messages, such as telephony, telecommunications services, and the internet.[6] The EAS is commonly used to alert the population to severe weather warnings, when evacuations may be ordered in response to flooding or HAZMAT emergencies, or to simply conduct tests to assure that the system is operational and capable of disseminating information when needed. More automated or sophisticated public alerting and warning systems are used in real time to issue warnings, such as using sirens that are activated when a tornado is approaching a community. Public address systems may be used in downtown business districts or on college campuses, where significant pedestrian traffic in public areas makes using loudspeakers to issue official news and directions a practical method. Some systems generate permanent records when they are used, while others do not.

Due to the wide reach of the EAS and other alerting and warning systems, and because their intended use is as an early-warning and mitigation system, the consequences of inaccurate information from such systems is likely to be significant. Such consequences can occur as a result of a variety of threats as well, including man-made deliberate and man-made unintentional threats. Man-made deliberate threats can include individuals who want to cause confusion or mayhem, and they may elect to perform their attack(s) during other incidents—to leverage it as a "force multiplier" of their own attack.

MAN-MADE UNINTENTIONAL THREAT: EMERGENCY MANAGEMENT

Unintentional threats can also cause such an incident, which would most likely occur as a consequence of carelessness or fatigue and result in an accidental release of inaccurate information. Unintentional releases of such information could be detected immediately, for example, by a joint information center, while deliberate attacks might not be as easy to detect because the deliberate actor may not want to be detected in the process of conducting the attack.

The vulnerabilities or enablers of a release of inaccurate or false information via an alerting and warning system could occur for a variety of reasons, mostly related to human and process vulnerabilities. Public announcements are typically vetted through an approval process that seeks to minimize the potential for release of false or inaccurate information. These procedures usually include review of the materials by more than one individual to ensure the accuracy of its content. In addition, these procedures can vary across various media. For instance, an EAS that has been in use for several years may have very mature and tested procedures. However, new technologies, such as online social media services, which have large user bases, have increasingly become useful tools for emergency management agencies because they are inexpensive to manage and maintain, and users can easily gain access to notifications from their local emergency management or emergency services entity/entities. The procedures for vetting information that goes out on these newer technologies are typically spontaneous in nature or do not readily allow for compliance with long-established processes. New technologies—because of their emphasis on having user-friendly interfaces—can also lend themselves to simple mistakes, such as an individual accidentally sending out a personal message to the emergency management agency's official account.[7]

Impacts of social media mishaps have historically resulted in embarrassment to the entity whose account was mistakenly used. For other alerting and warning systems, though, the public dissemination of inaccurate or incorrect information regarding a real or perceived incident could have a significant impact on local jurisdictions and their citizenry. The most immediate impact, unless the error is immediately detected, is that organizations and individuals can begin mobilizing or reacting to the information in ways that affect the community, including reallocation of emergency response, law-enforcement, and fire and emergency services personnel and resources, which would divert their limited resources from potentially legitimate and more urgent incidents or emergencies.

To manage the risks associated with a deliberate or unintentional release of inaccurate or false information from an alerting and warning systems, ESS entities should account for the tools and technologies that they use for the public notification of incidents and ensure that timely and sufficient policies and procedures are in place to manage the posting and release of information through them. Furthermore, the emergency management discipline, and ESS entities in general, should practice basic cybersecurity measures on a regular basis, while also identifying unique security enhancements that are necessary to manage the appropriate release of information through alerting and warning systems (e.g., regularly changing passwords and establishing complex passwords).

Scenario 5: Loss of Communications Lines Results in Disrupted Communications Capabilities

Scenario 5 focuses on loss as a result of man-made deliberate and man-made unintentional threats to all ESS-related communications. This scenario expands the scope of scenario 1. The components of scenario 5 include undesired consequences, the vulnerabilities that can lead to those undesired consequences, and the threats that can exploit those vulnerabilities. The components are relevant to all ESS disciplines but focus on the EMS, public works, and PSC&C disciplines.

Loss of communications is one of the most difficult scenarios that ESS faces. Every discipline, all elements of the value chain, and a majority of the supporting cyber infrastructure can be affected. As the ability to communicate is foundational to many pieces of the supporting cyber infrastructure, when communications systems fail, situational awareness for law-enforcement officers, firefighters, EMS providers, emergency managers, public works specialists, and the general public can be severely degraded. Depending on the exact nature of the situation, all parties will have to use less efficient and less effective ad hoc communications methods, such as public broadcasting, door-to-door notification, and word of mouth.

Risk Assessment Scenario 5: Loss of Communications Lines Results in Disrupted Communications Capabilities

In the case of a man-made deliberate threat in this scenario, the relative consequence is generally high and the likelihood of the threat is generally low, though medium specifically for fire and emergency services discipline. The ESS disciplines affected in this scenario will encounter roughly similar consequences depending on the specific communications lines lost.

For a man-made unintentional threat, the likelihood of the scenario is slightly higher for some disciplines, but the consequences are generally are lower. The likelihood will be slightly higher because there are a significant number of construction workers and other individuals who could potentially interfere with communications lines through carelessness or lack of awareness. The PSC&C, fire and emergency services, and EMS disciplines are especially affected in this case. The consequence will likely be slightly lower because mistakes may tend to be more minor because unintentional threat actors usually attempt to adhere to policy that prevents this scenario.

MAN-MADE DELIBERATE THREAT: FIRE AND EMERGENCY SERVICES

Deliberate acts that cause the loss of communications lines include the theft of copper wire from radio frequency (RF) infrastructure sites, eliminating the antenna's connectivity and/or the power supplies needed to support RF equipment and, in many cases, to carry radio traffic. These acts also include, but are not limited to, the theft of RF infrastructure components and vandalism to RF infrastructure sites. The result of these acts may not cause a widespread communications failure but are capable of causing local failures. The cumulative effect of a high number of such thefts and vandalism has made this a national-level threat.

As a consequence of such deliberate acts, fire and emergency services personnel may partially or completely lose connectivity between units or with dispatchers. Full or partial disruptions jeopardize the health and safety of firefighters, HAZMAT response teams, and others delivering service in this discipline because they may be unable to notify incident commanders when they urgently need additional resources or when they are lost, disoriented, or trapped in an environment that presents an immediate danger to life and health. These disruptions also jeopardize the general public because they may delay the deployment of firefighters to burning buildings or to HAZMAT spills. Law-enforcement personnel may also partially or completely lose connectivity with other units or with dispatchers. Such disruptions can decrease officer safety because they lose situational awareness of complaint responses, traffic stops, and suspicious persons and events that other officers are encountering. This may delay backup when it is most needed, which could place officers in the position of using greater force to overcome a threat to themselves or to the public than would otherwise be necessary, if additional units were on the scene. Law-enforcement officers encountering high-risk situations may be at greater risk of bodily harm if they cannot communicate with other officers or their dispatchers.

The greatest threat to the RF infrastructure from deliberate acts today lies with thieves. Scrap metal prices are at record highs because society has placed a high value on recycling metals rather than creating them from raw materials, and the cost to accomplish this and distribute recycled materials of high quality and quantity has reached commercially viable levels. Copper has been a particularly lucrative metal seen at scrap facilities. This sort of market makes taking the risks of discovery, prosecution, or even electrocution acceptable to thieves. One such example of the extreme risks that thieves are willing to take occurred in Dallas, Texas, in March 2010, when a group of young men were electrocuted and burned to death when they attempted to steal an energized (13,200 volts alternating current) copper line.

There are a few key vulnerabilities associated with these threats:

- Insufficient physical security around cyber assets. Many RF sites for land mobile radio (LMR) systems are located in remote areas based on coverage requirements. Therefore, they may be atop a mountain or perched on poles in a large field to carry signal between dispatchers and field forces. Some sites are located on private property, where agencies providing LMR services may not have the ability to erect the physical barriers they need.

Others may have limited barriers, such as simple fence lines, without any supporting services such as guards, surveillance cameras, or alarm systems, because of limited power supplies or connectivity.
- Easily identifiable and fairly accessible part of infrastructure. Copper wire, even when it is insulated, is easy to identify. Even thieves with limited experience with wire or electrical systems can easily discern copper wire at connection points. Often, copper lines are connected to RF infrastructure outdoors or in areas where it is obvious that copper resources are being used, such as along radio tower components.
- Dependence on third-party communications providers. Many agencies use commercial service providers to erect, maintain, and repair their LMR systems because it may be more economical than operating an internally staffed and equipped radio shop. Under such circumstances, it is more difficult to determine whether these providers are exercising sufficient security measures to assure that their staff members are provided with effective security policies and guidelines, are well trained, are trustworthy, and are reliable when servicing and protecting these systems, or whether they have implemented and maintained security measures that provide adequate protection for the infrastructure for which they care.

The general risk in this scenario indicates a moderately low likelihood that such deliberate acts will occur to any given fire and emergency services communications network. Rural agencies may face greater likelihood because their RF sites may have less physical security available; in suburban and urban areas, these sites are generally in areas within public view and have greater physical security measures protecting them. Nonetheless, the consequences of such deliberate acts can be equally grave, and they present the highest risk if they occur.

Although the deliberate act presents the greatest threat to communications line in either their “first mile” or “last mile” (i.e., when they are most visible to thieves and when their connections to critical infrastructure are most apparent), the unintentional loss of these lines almost always occurs when the lines are not visible. It is when the presence of communications lines is not known because they are buried, or because they are “out of sight and out of mind” when they are carried overhead. It is when the RF site components to which they are connected may not be recognized or understood for the essential support they provide because they are hidden away in a closet or they appear to be part of other electronic applications, such as computers or telephony. Likewise, telephone lines that support LMR communications may not be recognized when bundled with regular telephone service landlines. This relative invisibility creates the risk of an unintentional loss of local communications lines.

MAN-MADE UNINTENTIONAL THREAT: EMERGENCY MEDICAL SERVICES

Accidental disruptions may occur in a variety of circumstances. When the telephone lines connecting LMR sites with emergency communications centers are collocated with other telephone lines, they may be prone to accidental interruptions when disconnections, relocations, or other legitimate and planned outages are otherwise taking place. Construction projects and utility work for other interests create risk for these lines because such activities rely on accurate depictions on maps and well-placed ground markers and utility flags to avoid accidentally severing such lines while using earth-moving equipment. Disruption can be caused by something as simple as an incorrectly placed shovel or as complex as a backhoe striking a cable bundle in which communications lines are trunked or collocated with dozens, or perhaps even hundreds, of other lines. Communications lines can also be lost when critical RF components fail or are incorrectly manipulated. If a base interface module (BIM) fails because of a defect, a critical RF component fails because it has reached the end of its life cycle, or a dispatcher accidentally disables a repeater site from his or her radio console, the resulting disruptions to emergency communications are the same. While these disruptions can affect any ESS discipline, EMS providers may be particularly affected by accidental disruptions. The high volume of emergency calls to which this discipline responds requires reliable LMR communications to transmit and receive dispatches, make hospital notifications, seek online medical direction, and assure that the proper level of care providers are routed to incidents for which their skills are best suited.

The time to make repairs or corrections will vary widely in these various accidental disruptions. The accidental command or erroneous manipulation of controls can be quickly detected and corrected. If a BIM card fails, as long as a replacement or spare is readily available, the problem can be resolved in minutes if someone is available who is trained to recognize and troubleshoot that problem. The loss of communications lines somewhere between that first and last mile of connectivity is a much more difficult problem. It can take hours to locate the source of such trouble, depending on the cause, and hours or days to make repairs, depending on how many lines in the affected cable bundle were severed.

There are a few key vulnerabilities associated with these threats:

- Insufficient physical security around cyber assets. As noted before, the ability to secure emergency communications infrastructure presents a significant vulnerability that can be easily exploited, even if done so accidentally

or unintentionally. In the latter circumstance, it is the collocation of communications lines with other lines not used by ESS that presents vulnerability to accidental disruption. It is difficult, if not impractical, to protect miles of telephone lines leased to ESS organizations from the effects of weather, deterioration, or cutting in a manner separate from regular commercial or residential communications lines.
- Human factors may present unanticipated vulnerabilities resulting from accidents. Whether the people who may accidentally cause the loss of a communications line are part of an ESS organization or not, the interaction of people and their activities in proximity to essential communications lines is ever present. Construction projects or utility repairs occur across the country on most weekdays. Equipment operators may be well trained, but if the presence of underground utilities has not been determined, or if the markers, such as flags, have been altered by passersby or others, their earthmoving equipment can easily inflict permanent or long-term damage in a matter of moments. Communications staff, whether they are dispatchers or technicians, can accidentally disrupt these lines with errors in commands or by performing other acts that unwittingly disconnect lines, such as cleaning activities, moving equipment, or shutting off unrelated equipment.
- Variance in the capabilities of third-party providers in addressing communications disruptions. In most instances across the country, ESS must rely on commercial service providers for the connectivity among their proprietary LMR systems terminal points, such as RF sites, communications centers, and broadcast towers. There are alternatives available, such as microwave, to carry signals between RF sites, but in the end, a telephone is still a line used to carry that signal to the dispatch console or to the broadcast antenna. The ability of a commercial service provider to detect and repair a disruption can vary widely, based on factors such as technician availability, the age and condition of the infrastructure, weather, and even time of day or day of the week if a disruption occurs after regular business hours.

MAN-MADE UNINTENTIONAL THREAT: PUBLIC SAFETY COMMUNICATIONS AND COORDINATION/FUSION

Degraded emergency communications capabilities can lead to a number of impacts. These include:

- Inefficient or ineffective allocation of resources during an incident
- Longer dispatch and response times because of a lack of real-time or near real-time situational awareness of people, resources, and emergencies

- Incomplete or lack of alerting and warning capabilities used to indicate or announce emergencies to the general public.

These consequences can cascade into loss of life, loss of economic security, impacts on health-care and public health services, and widespread loss of confidence in government, its services, or its messages to the public.

Telecommunications networks and infrastructure are critical to the PSC&C discipline because first responders and public safety communications agencies use a variety of communications services, from POTS to cellular telephones and smartphones, pagers, and personal digital assistants (PDAs). These services help public safety agencies receive and process calls, analyze location data about callers, and make notifications to allied services (e.g., law enforcement, fire and rescue, public utilities). Public safety radio networks (such as LMR and radios in the aviation and marine bands that engage in patrol, response, search, rescue, and evacuation operations) use telecommunications networks for backhaul services. In addition, reliance on telecommunications services is increasing steadily as more and more public safety agencies transition from legacy analog services.

The primary threats that could cause this scenario are man-made unintentional threat actors who could accidentally damage or sever communications lines.[8] These include employees, third-party contractors, construction workers, maintenance workers, and telecommunications service workers. With each of these actors, the action could be caused by carelessness/recklessness (e.g., not coordinating with utility location services to understand where communications lines are located before starting construction), a lack of training, or a technical flaw. The vulnerabilities include a lack of route diversity, aging infrastructure that is not well protected, mismarked utility markings, or a lack of attention paid to utility markings. Overall, the likelihood of these threats with respect to these vulnerabilities was determined to be low to medium.

Direct consequences of this scenario to the ESS in general and the PSC&C discipline in particular may include the loss of the HPH sector's ability to provide adequate care and emergency response. With the dependence of all ESS disciplines on the PSC&C discipline, the consequences were assessed to be medium to high, depending on the severity and duration of the outage. The impact of this scenario is likely to be felt more strongly on a local scale. At a regional level or higher, there is significantly more redundancy and route diversity built into the network, resulting in greater failover capabilities, resilience, and service continuity. Moreover, the consequences of this scenario may be mitigated on even a local level by increased route diversity or better access controls. However, it is important to note that in this scenario, recovery and reconstitution of the network depends primarily on third-party communications service providers and are therefore largely out of the control of public safety agencies. This also means that public safety organizations have fewer options to directly mitigate or reduce key vulnerabilities that could cause this scenario. Furthermore, while the impact may be felt only at a local level, it may put significant strain on local public safety resources.

Scenario 6: Closed-Circuit Television Jamming/Blocking Results in Disrupted Surveillance Capabilities

Closed-circuit television (CCTV) jamming/blocking that results in disrupted surveillance capabilities would most likely affect the public works, emergency management, and law-enforcement disciplines. Many CCTV networks are switching to IP-based communications, creating new vulnerabilities for threat actors to exploit. Older CCTV networks are also prone to attacks from various threats. Potential consequences from CCTV jamming/blocking include the inability of law-enforcement personnel to apprehend criminals, difficulties for emergency management personnel in trying to identify where to allocate resources, potential public panic or chaos if traffic systems are affected, and disruption to public works' ability to monitor and/or respond to incidents. The components of this scenario include undesired consequences, the vulnerabilities that can lead to those undesired consequences, and the threats that can exploit those vulnerabilities. The components are relevant to the following ESS disciplines: law enforcement, EMS, emergency management, and public works.

In the past decade, CCTV has evolved into a highly efficient and effective public safety tool used to deter crime, monitor critical assets, and provide the technology needed to manage public safety responses in real time. When CCTV is jammed or blocked, law enforcement and emergency managers lose their "eyes and ears," thereby degrading their ability to monitor critical assets (e.g., buildings, power plants, roads, airports, bridges) as well as to monitor and manage public safety events. In these types of situations, monitoring and surveillance may need to be performed by placing law-enforcement officers physically on-site.

Risk Assessment Scenario 6: Closed-Circuit Television Jamming/Blocking Results in Disrupted Surveillance Capabilities

In the case of a man-made deliberate threat in this scenario, the likelihood of the threat affecting the ESS disciplines is low to medium, and the relative consequences range from high for public works, medium for emergency management and law enforcement, and low for EMS. The public works discipline in general will be highly affected because of its need to monitor locations to accomplish its mission. Because the scope of consequences for this scenario is mostly local, the relative consequences to the disciplines will vary depending on the target of the man-made deliberate threat.

MAN-MADE DELIBERATE THREAT: PUBLIC WORKS

The jamming or blocking of CCTV systems hinders the public works discipline's ability to monitor critical infrastructure facilities and systems or high-risk infrastructure areas. The inability to monitor these facilities and areas impedes public works agencies from detecting and preventing physical incidents. The situational awareness provided by CCTV also assists ESS in responding to and recovering from physical incidents such as motor vehicle collisions or disabled vehicles blocking traffic. In the event of a jammed or blocked CCTV, this response could be delayed. Other potential incidents monitored by CCTV cameras include debris blockages in wastewater processing systems and mechanical failures in other environmental systems. The general public will likely be affected through increased traffic congestion or possibly through the disruption of utility services. The effects of a jammed or blocked CCTV system could cascade to other sectors, including the dams, water, and transportation systems sectors, which have close ties at the local level to the public works discipline.

Cyber vandals are the typical threat actors associated with this type of incident, although any actor with a physical component to its objectives may attempt to jam or block a CCTV system to mask its activities. Cyber vandals often target CCTV systems for their own amusement or to cause embarrassment to public works officials. These types of attacks require minimal funding and are often crimes of opportunity rather than intricately planned endeavors. In most cases, tools, such as signal jammers, are required, but these are readily available and easily adapted to fit the actor's needs. These types of actors often seek to keep their identities hidden and rarely commit more than minor criminal acts, such as vandalism or trespassing. In addition, actors often must gain physical or logical access to the CCTV system, which increases the difficulty of jamming or blocking the system. Physical security measures and placing cameras out of reach further constrain an actor's ability to gain physical access to the system, and system monitoring and access controls can constrain logical access to the system. Enhanced redundancy measures, such as employing multiple cameras, reduces the overall risk posed by a jammed or blocked CCTV system.

Several vulnerabilities exist within the people, processes, and technology associated with CCTV systems that can be exploited by threat actors to jam or block the system. CCTV control stations placed in public spaces increase the time a threat actor has to access the system. Employees may knowingly or unknowingly disclose the location of control stations or cameras to threat actors, or even execute commands on behalf of the threat actor. In some cases, these vulnerabilities are introduced as a result of inadequate clearance checks and employee vetting, a lack of operational security, or a lack of communications security related to CCTV system use. The technology itself is vulnerable to exploitation, and the move to IP-based systems provides threat actors with the ability to use cyber exploits to jam or block CCTV systems and retrieve or manipulate the audio and video recordings stored on those systems. Measures currently employed by public works agencies to address these types of vulnerabilities include placing control stations in secured facilities, performing rigorous employee vetting and training, and implementing industry standards and guidelines for the management of CCTV systems.

The jamming or blocking of a CCTV system would lead to significant undesired consequences for the public works discipline. However, there is a low likelihood that this type of incident will occur because of the risk responses currently employed by the discipline. In the future, risks to the discipline's CCTV systems could change or increase as public works agencies move to IP-based CCTV systems.

Scenario 7: Overloaded Communications Network Results in Denial-of-Service Conditions for Public Safety and Emergency Services Communications Networks

This scenario specifically focuses on the loss of availability of PSC&C networks as a result of denial-of-service conditions. This scenario can occur deliberately as a result of a malicious actor launching a denial-of-service attack or unintentionally as a result of a network overload caused by a sudden and unexpected surge in public use. In addition to affecting wireless communications networks, this scenario is applicable to Next Generation 911 and IP-based and other cloud-based communications networks, which are also vulnerable to man-made deliberate and unintentional threats and are of increasing importance in ESS.

The components of this scenario include undesired consequences, the vulnerabilities that can lead to those undesired consequences, and the threats that can exploit those vulnerabilities. The components are relevant to all ESS disciplines but focus on the law-enforcement, fire and emergency services, EMS, and PSC&C disciplines. An overloaded communications system or network that results in a denial of service could affect multiple disciplines, segments of the value chain, and numerous supporting cyber infrastructure components, particularly those components that are IP based. In this scenario, situational awareness tools such as CAD and GIS tools could be negatively affected across the value chain, potentially resulting in a life-threatening delay of service to customers. In the worst cases, typical ESS services would be unavailable to customers until the issues are resolved. A compounding factor of this type of scenario is that such an overload is likely the result of a catastrophic event, creating a situation in which the system is unresponsive or slow to respond and emergency services agencies cannot be reached when the most people need help, causing potential loss of life and damage to property.

Risk Assessment Scenario 7: Overloaded Communications Network Results in Denial-of-Service Conditions for Public Safety and Emergency Services Communications Networks

This scenario affects all of the ESS disciplines. In the case of a man-made deliberate threat in this scenario, the likelihood of the threat affecting the ESS disciplines is low, but the consequences for all disciplines are medium to high. The likelihood is low because of the effort level associated with deliberately triggering the scenario. A man-made deliberate actor would need to focus significant resources on one location over a period of time to successfully execute this scenario. However, given the increasing strength of botnets (collections of compromised computers connected to the internet) available to hackers, this attack may be more feasible in the near future, especially for localized areas.[9]

For a man-made unintentional threat, the likelihood for the fire and emergency services and EMS disciplines increases to medium. An example of a common man-made unintentional threat for this scenario would be a large event or gathering. The large event or gathering would result in a large volume across local communications networks, thus creating an unintentional distributed denial of service.

MAN-MADE DELIBERATE THREAT: EMERGENCY MEDICAL SERVICES AND FIRE AND EMERGENCY SERVICES

EMS and fire and emergency services agencies use a variety of wireless cyber resources across the country. Many wireless networks are being adopted for use in supporting emergency, nonemergency, and administrative communications requirements for voice, data, and in some cases video services. In most parts of the country, emergency communications are conducted using proprietary LMR networks and dedicated private lines for landline connectivity between communications centers and hospitals, aeromedical evacuation services, law enforcement, and fire and emergency services agencies. When other wireless services are used, they are generally the same services used by the average American consumer—enhanced special mobile radio (ESMR), cellular telephones, short-message services, and data. When the demand for service by ESS and the public peak simultaneously, the networks supporting these commonly accessible services are overloaded. The result is that communications can deteriorate significantly.

Major emergencies, such as the terrorist attacks of September 11, 2001, have proven that even in urban areas in which commercial communications services are robust and built to serve hundreds of thousands of call requests in a short period of time, the availability of these systems can be lost in less than an hour. On August 23, 2011, a moderate earthquake centered near Mineral, Virginia, generated a huge demand for wireless communications access in just a few moments. The available infrastructure was unable to support these demands and resulted in busy signals or lost calls in less than ten minutes in such population centers as Washington, DC—more than ninety miles from the epicenter of the quake. A government program designed to provide access to critical end users in government and emergency services, the Wireless Priority Service (WPS), failed to support authorized subscribers on any commercial systems.[10]

When a surge in demand can be predicted, commercial service providers can prepare by placing extra temporary infrastructure into position. For example, in 2009, the District of Columbia Homeland Security and Emergency Management Agency developed a Presidential Inauguration Communications Plan in which it made assumptions about record demands for wireless communications services as more than one million Americans were believed to be planning to attend the historic inauguration of the country's first African American president. The agency worked with commercial communications providers to address their concerns. In a regional after-action discussion, National Capitol Region ESS representatives noted that three major service providers placed cell-on-wheels and cell-on-light-truck units at a number of downtown locations, an action that proved itself invaluable as demands spiked at more than one thousand times greater than normal while Barack Obama took his oath of office. In the vicinity of the National Mall, people attending the event sent text messages, held their mobile telephones with open lines to loudspeakers, or sent photos to share their experience with others who could not be there. Although many people experienced dropped calls or were denied access to their networks because of the volume in demand, the systems did not crash. More important, in that instance, the WPS worked when its subscribers, including local fire and emergency services officials, were unable to complete calls using conventional dialing.

Given the examples discussed here, it is apparent that any incident that creates a widespread concern among the public and a subsequent surge in demand for wireless communications access can create an overloaded network. The ability to predict such surges, such as for scheduled events, provides commercial service providers with an opportunity to apply temporary infrastructure to help meet that surge. It is when the unpredictable event or emergency incident occurs that an overloaded network presents the greatest threat to ESS.

There seem to be three major consequences for EMS agencies when wireless communications networks are overloaded:

- Loss or degradation of 911 and other emergency mobile communications. EMS organizations may use such commercial services as ESMR to coordinate activities with other government agencies serving in their area or to conduct administrative communications with communications centers or firehouses. They may use cellular telephones to communicate with outside resources such as language-line services, CHEMTREC 17, poison control, or online medical control when they require subject-matter expertise to effectively manage a response problem. EMS may use cellular telephones to obtain authorization to implement medical interventions that exceed standard protocol to save a seriously ill or injured patient. As the public is competing for access to these same commercial networks, callers may experience degraded abilities to call for help because they are unable to complete or lose calls placed to 911 centers that support the EMS and fire and emergency services agencies.
- Inability to deploy resources effectively. Although the EMS discipline may not use commercial wireless networks for emergency communications (e.g., LMR support), it may use commercial networks to support key resources such as mobile data networks for dispatching and deploying field forces, jurisdiction-specific navigational aids, or access to information services such as online medical libraries or building preplans. This can cause delays or lost assignments, or lost access to key medical information at a time when lives are at stake.
- Confusion/panic and loss of public confidence in emergency services. When one is in dire need of EMS or fire and emergency services, the ability to acquire a dial tone and complete a call to 911 or other emergency telephone numbers is essential. When callers cannot reach anyone by dialing an emergency number, it creates doubt about the effectiveness of the response system. People become confused because they may not be aware of other numbers to dial for help, or they may panic because they do not know how to summon the help they need. Many people, especially travelers, have no idea at any given time where the nearest hospital, rescue squad, or fire station may be, and so they cannot even make a personal report of an emergency to get help started.

Fortunately, the likelihood of an emergency significant enough in scope to create an overloaded network is low for any given EMS or fire and emergency services agency. Although there are hundreds of thousands of emergencies that these disciplines respond to every day, the number of those incidents that escalate to create such surges are very few. Therefore, while the consequences, should such an incident occur, are very high, the likelihood is low.

NOTES

Much of this chapter is adapted from the U.S. Department of Homeland Security Cybersecurity and Infrastructure Security Agency's 2012 report titled *Emergency Services Sector Cyber Risk Assessment* (https://www.cisa.gov/resources-tools/resources/emergency-services-sector-cyber-risk-assessment).

1. "The Disaster Process and Disaster Aid Programs," http://www.fema.gov/hazard/dproc.shtm.
2. Multiple types of attacks against databases are included in the 2010 Open Web Application Security Project's "Top 10 Critical Web Application Security Flaws," https://www.owasp.org/index.php/Top_10_2010.
3. SANS, "Top Cyber Security Risks," http://www.sans.org/top-cyber-security-risks/summary.php.
4. NEMSIS, http://www.nemsis.org.
5. NEMSIS, http://www.nemsis.org.
6. An example of this is a reverse 911 system, which can be used to rapidly dial the home and business telephones in a defined segment of a community, an entire community, or even an entire jurisdiction to issue warnings such as "boil water orders" when a major public water distribution failure has occurred. Local, state, federal, and tribal emergency management agencies have some level of access to the EAS, whether it is with local television or radio broadcast stations or satellite and cable communications companies, or other methods of access.
7. For an example of mistaken use of a corporate social media account instead of a personal one, see https://money.cnn.com/2011/02/17/smallbusiness/dogfish_redcross/index.htm.
8. It should also be noted that while this scenario focuses on man-made deliberate or unintentional threats to communications lines, natural disasters can also cause the same or similar effects as those expressed in this scenario's analysis. Since natural disasters are specifically discussed in scenario 1, the participants did not evaluate the impact of a natural disaster on physical communications lines.
9. The Federal Communications Commission's Communications Security, Reliability, and Interoperability Council III has a working group dedicated to developing recommendations and guidance for botnet remediation (http://www.fcc.gov/encyclopedia/communications-security-reliability-and-interoperability-council-iii).
10. As reported by the online magazine *NextGov*, available at http://www.nextgov.com/nextgov/ng_20111128_2122.php?oref=search.

Chapter 7

Emergency Services Sector Contingency Planning

Emergency response planning or contingency planning for extreme events has long been standard practice for emergency planners and safety professionals in industrial systems operations. For many years, prudent practices have required consideration of the potential impact of severe natural events (forces of nature), including earthquakes, tornadoes, volcanoes, floods, hurricanes, and blizzards. These possibilities have been included in ESS infrastructure emergency preparedness and disaster response planning. In addition, many ESS service facilities have considered the potential consequences of man-made disasters such as those resulting from operator error and manufacturer defects in industrial equipment. Currently, ESS managers and operators must also consider violence in the workplace. Moreover, at present, as this text has pointed out, there is a new focus of concern: the potential effects of intentional acts by domestic (homegrown, in-house) or international (foreign) terrorists.

As a result, the security paradigm has not necessarily changed but instead has been radically adjusted—reasonable, necessary, and sensible accommodation for and mitigation of just about any emergency situation imaginable have been and continue to be practiced. Because we cannot foresee all future domestic or intentional acts of terrorism, we must be prepared to shift from the proactive to the reactive mode on short notice—in some cases, on very short notice. Accordingly, we must be prepared to respond to and mitigate what we cannot prevent. Unfortunately, there is more we can't prevent than we can prevent. In light of this, in this section we present in outline form a reactive mitigation procedure, the template example for a standard ESS crisis plan, dealing with natural and man-made disasters, which could also be applied in response to acts of terrorism.

CRISIS EMERGENCIES SERVICES PLAN

Note: The following criteria have not been established as anything other than guidelines and are offered not as definitive or official regulations or procedures but rather as informed advice (based on more than thirty years of safety, industrial hygiene, emergency contingency planning, and security experience) for both the public and private sectors.

Note: This emergency action plan applies to locations and facilities that are occupied by emergency services personnel performing their designated work activities.

The fact is well known: when an emergency occurs, the need to communicate is immediate. The goals of an emergency services plan are to document and understand the steps needed to:

- Rapidly restore material and machinery producing, refining, transporting, generating, transmitting, conserving, building, distributing, maintaining, and controlling systems and components after an emergency
- Minimize damage to communications production process equipment
- Minimize impact and loss to customers
- Minimize negative impacts on public health and employee safety
- Minimize adverse effects on the environment
- Provide emergency public information concerning customer service
- Provide hazardous chemical information for first responders and other outside agencies

Although we are concerned with the ESS in this text, the EPA-developed *Large Water System Emergency Response Outline: Guidance to Assist Community Water Systems in Complying with the Public Health and Bioterrorism Preparedness and Response Act of 2002* (dated July 2003) has been successfully implemented and tested, and with minor adjustments can be applied as a template for the ESS. This template provides guidance and recommendations to aid facilities in the preparation of emergency response plans (ERPs) under Public Law 107-188. The template is provided below.

EMERGENCY SERVICES SECTOR PLAN TEMPLATE

I. Introduction

Safe and reliable operation is vital to every industrial, administrative, and service operation. An emergency services plan (ESP) is an essential part of managing a critical infrastructure unit, process, or entity. The introduction should identify the requirement to have a documented ESS, the goal(s) of the plan (e.g., be able to quickly identify an emergency and initiate timely and effective response action, be able to quickly respond and repair damages to minimize system downtime), and how access to the plan is limited. Plans should be numbered for control. Recipients should sign and date a statement that includes their (1) ESP number, (2) agreement not to reproduce the ESP, and (3) acknowledgment that they have read the ESP.

ESPs do not necessarily need to be one document. They may consist of an overview document, individual emergency action procedures, checklists, additions to existing operations manuals, appendices, etc. There may be separate, more detailed plans for specific incidents. There may be plans that do not include particularly sensitive information and those that do. Existing applicable documents should be referenced in the ESP.

II. Emergency Planning Process

A. Planning Partnerships

The planning process should include those parties who will need to help the emergency services sector in an emergency situation (e.g., first responders, law enforcement, public health officials, nearby utilities, local emergency planning committees, testing labs, etc.). Partnerships should track from the ESS operation up through local, state, regional, and federal agencies, as applicable and appropriate, and could also document compliance with governmental requirements.

B. General Emergency Response Policies, Procedures, Actions, and Documents

A short synopsis of the overall emergency management structure, how other industrial emergency response, contingency, and risk management plans fit into the CMP for ESS emergencies, and applicable policies, procedures, actions plans, and reference documents should be cited. Policies should include interconnected agreements with adjacent communities and just how the CMP may affect them.

C. Scenarios

Use your vulnerability assessment (VA) findings to identify specific emergency action steps required for response, recovery, and remediation for applicable incident types. In section V of this plan, specific emergency action procedures addressing each of the incident types should be addressed.

III. Emergency Response Plan—Policies

A. System Specific Information

In an emergency, ESS operations need to have basic information for system personnel and external parties such as law enforcement, emergency responders, repair contractors/vendors, the media, and others. The information needs to be clearly formatted and readily accessible so system staff can find and distribute it quickly to those who may be involved in responding to the emergency. Basic information that may be presented in the ERP are the system's ID number, system name, system address or location, directions to the system, population served, number of service connections, system owner, and information about the person in charge of managing the emergency. Distribution maps, detailed plant drawings, site plans, source/storage/production energy locations, and operations manuals may be attached to this plan as appendices or referenced.

1. PWS ID, owner, contact person
2. Population served and service connections
3. System components
   a) Conduits and constructed conveyances
   b) Physical barriers
   c) Electronic, computer, or other automated systems that are utilized by the ESS
   d) Emergency power generators (on-site and portable)
   e) The operation and maintenance of such system components

B. Chain-of-Command Chart Developed in Coordination with Local Emergency Planning Committee (internal and/or external emergency responders, or both)

1. Contact name
2. Organization and emergency response responsibility
3. Telephone number(s) (hardwire, cell phones, faxes, email)
4. State twenty-four-hour emergency communications center telephone

C. Communications Procedures: Who, What, When

During most emergencies, it will be necessary to quickly notify a variety of parties both internal and external to the communications sector entity. Using the chain-of-command chart and all appropriate personnel from the lists below, indicate who activates the plan, the order in which notification occurs, and the members of the emergency response team. All contact information should be available for routine updating and readily available. The following lists are not intended to be all inclusive—they should be adapted to your specific needs.

1. Internal Notification Lists
   a) Operations dispatch
   b) ESS manager
   c) ESS processing manager
   d) Communications production/storage/distribution manager
   e) Facility managers
   f) Chief energy engineer
   g) Director of engineering
   h) Data (IT) manager
   i) Maintenance manager
   j) Other
2. Local Notification
   a) Head of local government (i.e., mayor, city manager, chairman of the board, etc.)
   b) Public safety officials—fire, local law enforcement (LLE), police, EMS, safety. If a malevolent act is suspected, LLE should be immediately notified and in turn will notify the FBI, if required. The FBI is the primary agency for investigating sabotage.
   c) Other government entities: health, schools, parks, finance, electric, etc.
3. External Notification Lists
   a) State department of environmental quality (DEQ)
   b) USEPA/USDOE/DHS/FCC
   c) State police
   d) State health department (lab)
   e) Critical customers (special considerations for hospitals; federal, state, and country government centers; etc.)
   f) Service aid
   g) Mutual aid
   h) Response partners not previously notified
4. Public/Media Notification: When and How to Communicate
   Effective communications are a key element of emergency response, and a media or communications plan is essential to good communications. Be prepared by organizing basic facts about the crisis and your chemical system. Develop key messages to use with the media that are clear, brief, and accurate. Make sure your messages are carefully planned and have been coordinated with local and state officials. Considerations should be given to establishing protocols for both field and office staff to respectfully defer questions to the utility spokesperson.

Be prepared to list geographic boundaries of the affected area (e.g., west of highway A, east of highway B, north of highway C, and south of highway D) to ensure that the public clearly understands the system boundaries.

E. Personnel Safety

This should provide direction as to how operations staff, emergency responders, and the public should respond to a potential toxic chemical release, including facility evacuation, personnel accountability, proper personal protective equipment as dictated by the risk management program and the process safety management plan, and whether the nearby public should be "in-place sheltered" or evacuated.

F. Equipment

The ERP should identify equipment that can obviate or significantly lessen the impact of terrorist attacks or other intentional actions on the public health and protect the safety and supply of communities and individuals. The ESS facility should maintain an updated inventory of current equipment and repair parts for normal maintenance work.

Because of the potential for extensive or catastrophic damage that could result from a malevolent act, additional equipment sources should be identified for the acquisition and installation of equipment and repair parts in excess of normal use. A certain number of "long-lead" procurement equipment should be inventoried and the vendor information for such unique and critical equipment maintained. In addition, mutual-aid agreements with other industries, and the equipment available under the agreement, should be addressed. Inventories of current equipment, repair parts, and associated vendors should be indicated.

G. Property Protection

A determination should be made as to what ESS responding and processing operations/facilities should be immediately "locked down," specific access control procedures implemented, initial security perimeters established, and possible secondary malevolent events considered. The initial act may be a diversionary act.

H. Training, Exercises, and Drills

Emergency response training is essential. The purpose of the training program is to inform employees of what is expected of them during an emergency situation. The level of training on a CMP directly affects how well an ESS facility's employees can respond to an emergency. This may take the form of orientation scenarios, tabletop workshops, functional exercises, etc.

I. Assessment
To evaluate the overall CMP's effectiveness and to ensure that procedures and practices developed under the CMP are adequate and are being implemented, critical communications sector industry staff should audit the program on a periodic basis.

IV. Emergency Action Procedures (EAPs)
These are detailed procedures used in the event of an operation emergency or malevolent act. EAPs may be applicable across many different emergencies and are typically common-core elements of the overall municipality ERP (e.g., responsibilities, notifications lists, security procedures, etc.) and can be referenced.

A. Event classification/severity of emergency
B. Responsibilities of emergency director
C. Responsibilities of incident commander
D. Emergency operations center (EOC) activation
E. Division internal communications and reporting
F. External communications and notifications
G. Emergency telephone list (division internal contacts)
H. Emergency telephone list (off-site responders, agencies, state twenty-four-hour emergency phone number, and others to be notified)
I. Mutual aid agreements
J. Contact list of available emergency contractor services/equipment
K. Emergency equipment list (including inventory for each facility)
L. Security and access control during emergencies
M. Facility evacuation and lockdown and personnel accountability
N. Treatment and transport of injured personnel (including electrocution and petrochemical exposure)
O. Petrochemical records—to compare against historical results for baseline
P. List of available labs for emergency use
Q. Emergency sampling and analysis (petrochemical)
R. Water use restrictions during emergencies
S. Alternate temporary chemical supplies during emergencies
T. Isolation plans for chemical supply, treatment, storage, and distribution systems
U. Mitigation plans for neutralizing, flushing, and collecting spilled chemicals
V. Protection of vital records during emergencies
W. Record keeping and reporting (FCC, FEMA, DHS, DOT, OSHA, EPA, and other requirements) (it is important to maintain accurate financial records of expenses associated with the emergency event for possible federal reimbursement)

X. Emergency program training, drills, and tabletop exercises
Y. Assessment of emergency management plan and procedures
Z. Crime scene preservation training and plans
AA. Communication Plans:
  1. Police
  2. Fire
  3. Local government
  4. Media
  5. And so forth
BB. Administration and logistics, including EOC when established
CC. Equipment needs/maintenance of equipment
DD. Recovery and restoration of operations
EE. Emergency event closeout and recovery

V. Incident-Specific Emergency Action Procedures

Incident-specific EAPs are action procedures that identify specific steps in responding to an operational emergency or malevolent act.

A. General response to terrorist threats (other than bomb threats and incident-specific threats)
B. Incident-specific response to man-made or technological emergencies
  1. Contamination event (articulated threat with unspecified materials)
  2. Contamination threat at a major event
  3. Notification from health officials of potential contamination
  4. Intrusion through supervisory control and data acquisition (SCADA)
C. Significant structural damage resulting from intentional act
D. Customer complaints
E. Severe weather response (snow, ice, temperature, lightning)
F. Flood response
G. Hurricane and/or tornado response
H. Fire response
I. Explosion response
J. Major vehicle accident response
K. Electrical power outage response
L. Water supply interruption response
M. Transportation accident response—barge, plane, train, semitrailer/tanker
N. Contaminated/tampered with water treatment chemicals
O. Earthquake response
P. Disgruntled employee response (i.e., workplace violence)
Q. Vandal response
R. Bomb threat response

- S. Civil disturbance/riot/strike
- T. Armed intruder response
- U. Suspicious mail handling and reporting
- V. Hazardous chemical spill release response (including material safety data sheets)
- W. Cybersecurity/SCADA system attack response (other than incident specific, e.g., hacker)

VI. Next Steps
- A. Plan review and approval
- B. Practice and plan to update (as necessary, once every year recommended)
  1. Training requirements
  2. Who is responsible for conducting training, exercises, and emergency drills?
  3. Update and assessment requirements
  4. Incident-specific requirements

VII. Annexes
- A. Facility and location information
  1. Facility maps
  2. Facility drawings
  3. Facility descriptions/layout
  4. Etc.

VIII. References and Links
- A. Department of Homeland Security: https://www.dhs.gov
- B. Environmental Protection Agency: https://www.epa.gov
- C. Federal Emergency Management Agency: https://www.fema.gov
- D. Local emergency planning committees: http://www.epa.gov/ceppo/epclist.htm

THE BOTTOM LINE

Because industrial emergencies (in less-than-extreme conditions) can seriously affect the surrounding community and environment, and because poor planning and/or panic can only make a bad situation worse, and can also lead to additional injury and death, your role as an ESS team or site manager or site safety professional in an emergency response is doubly important. A crisis out of hand can easily devastate a community—and your organization is (or should be) an active member of your community. By allowing a less-than-effective emergency response, communications site managers endanger not only themselves and their organizations but also their organization's community and standing as well.

REFERENCES AND RECOMMENDED READING

Brauer, R. L. (1994). *Safety and Health for Engineers.* New York: Van Nostrand Reinhold.

CoVan, J. (1995). *Safety Engineering*. New York: Wiley.

Healy, R. J. (1969). *Emergency and Disaster Planning*. New York: Wiley.

Federal Emergency Management Agency. (1981). *Planning Guide and Checklist for Hazardous Materials Contingency Plans*. FEMA-10. Washington, DC, July.

Office of the Federal Register. (1987). 29 CFR 1910.120. Washington, DC.

Smith, A. J. (1980). *Managing Hazardous Substances Accidents*. New York: McGraw-Hill.

Spellman, F. R. (1997). *A Guide to Compliance for Process Safety Management Planning/Risk Management Planning (PSM/RMP)*. Lancaster, PA: Technomic Publishing.

U.S. Army Corps of Engineers. (1987). *Safety and Health Requirements Manual*. Rev. ed. EM 385-1-1. Washington, DC, October.

USDOE. (2008). *Emergency Support Function #12—Energy Annex*. Washington, DC: U.S. Department of Energy.

USDOE. (2010). *Energy Sector-Specific Plan: An Annex to the National Infrastructure Protection Plan.* Washington, DC: U.S. Department of Energy.

USEPA. (2002). *Water Utility Response, Recovery & Remediation Guidance for Man-Made and/or Technological Emergencies*. Washington, DC: U.S. Environmental Protection Agency.

USEPA. (2003). *Large Water System Emergency Response Plan Outline: Guidance to Assist Community Water Systems in Complying with the Public Health Security and Bioterrorism Preparedness and Response Act of 2002*. EPA 810-F-03-007. Washington, DC: U.S. Environmental Protection Agency. https://19january2017snapshot.epa.gov/sites/production/files/2015-03/documents/erp-long-outline.pdf, accessed May 17, 2023.

Chapter 8

Security Techniques and Hardware

THE MULTIPLE-BARRIER APPROACH

Ideally, in a perfect world, all ESS physical sites/facilities/towers would be secured in a layered fashion (the "multiple barrier approach"). Layered security systems are vital. Using the "protection-in-depth" principle, requiring that an adversary defeat several protective barriers or security layers to accomplish its goal, ESS physical infrastructure can be made more secure—to a degree, that is. Obviously, because fire and rescue services need to respond to an alarm quickly, layered protection could impede their ability to respond immediately. Protection in depth is a term commonly used by the military to describe security measures that reinforce one another, masking the defense mechanisms from the view of intruders and allowing the defender time to respond to an intrusion or attack.

A prime example of the use of the multi-barrier approach to ensure security and safety is demonstrated by the practices of the bottled water industry. In the aftermath of 9/11 and the increased emphasis on homeland security, a shifted paradigm of national security and vulnerability awareness has emerged. Recall that in the immediate aftermath of the 9/11 tragedies, emergency responders and others responded quickly and worked to exhaustion. In addition to the emergency responders, bottled water companies responded immediately by donating several million bottles of water to the crews at the crash sites in New York City, at the Pentagon, and in Pennsylvania. The International Bottled Water Association (IBWA, 2004) reports that, "within hours of the first attack, bottled water was delivered where it mattered most; to emergency personnel on the scene who required ample water to stay hydrated as they worked to rescue victims and clean up debris" (p. 2).

Bottled water companies continued to provide bottled water to responders and rescuers at the 9/11 sites throughout the post-event processes. These patriotic actions by the bottled water companies, however, beg the question: How do we ensure the safety and security of the bottled water provided to anyone? The IBWA (2004) has the answer: using a multi-barrier approach, along with other principles, will enhance the safety and security of bottled water. The IBWA describes its multi-barrier approach as follows:

> A multi-barrier approach—Bottled water products are produced utilizing a multi-barrier approach, from source to finished product, that helps prevent possible harmful contaminants (physical, chemical or microbiological) from adulterating the finished product as well as storage, production, and transportation equipment. Measures in a multi-barrier approach may include source protection, source monitoring, reverse osmosis, distillation, filtration, ozonation or ultraviolet (UV) light. Many of the steps in a multi-barrier system may be effective in safeguarding bottled water from microbiological and other contamination. Piping in and out of plants, as well as storage silos and water tankers, are also protected and maintained through sanitation procedures. In addition, bottled water products are bottled in a controlled, sanitary environment to prevent contamination during the filling operation. (p. 3)

In ESS infrastructure security, where practicable, protection in depth is used to describe a layered security approach. A protection-in-depth strategy uses several forms of security techniques and/or devices against an intruder and does not rely on one single defensive mechanism to protect infrastructure. By implementing multiple layers of security, a hole or flaw in one layer is covered by the other layers. An intruder will have to penetrate through each layer without being detected in the process—the layered approach implies that no matter how an intruder attempts to accomplish his goal, he will encounter effective elements of the physical protection system.

For example, as depicted in figure 8.1, an effective security-layering approach requires that an adversary penetrate multiple, separate barriers to gain entry to a critical target at an ESS facility. As shown in figure 8.1, protection in depth (multiple layers of security) helps to ensure that the security system remains effective in the event of a failure or an intruder bypassing a single layer of security.

Again, as shown in figure 8.1, layered security starts with the outer perimeter (the fence—the first line of physical security) of the facility and goes inward to the facility—the buildings, structures, and other individual assets—and finally to the contents of those buildings: the targets.

The area between the outer perimeter and the structures or buildings is known as the site. This open site area provides an incomparable opportunity for early identification of an unauthorized intruder and initiation of early

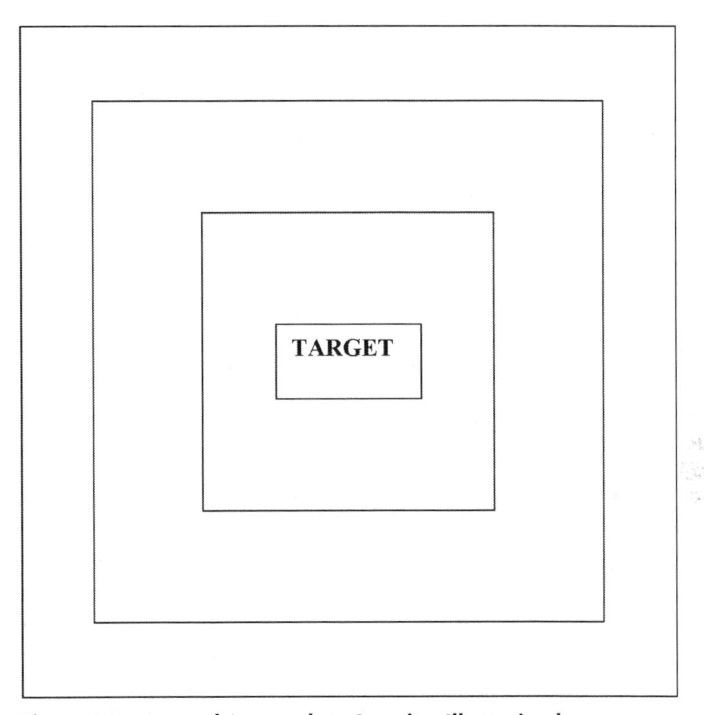

Figure 8.1. Layered Approach to Security. ***Illustration by F. R. Spellman and K. Welsh***

warning/response. This open space area is commonly used to calculate the standoff distance; that is, it is the distance between the outside perimeter (public areas to the fence) to the target or critical assets (buildings/structures) inside the perimeter (inside the fence line—the restricted access area).

The open area, between perimeter fence and target (e.g., the operations center), if properly outfitted with various security devices, can also provide layered protection against intruders. For example, lighting is a deterrent. Based on personal experience, an open area within the plant site that is almost as well lighted at night as would be expected during daylight hours is the rule of thumb. In addition, strategically placed motion detectors along with crash barriers at perimeter gate openings and in front of vital structures are also recommended. Armed, mobile guards who roam the interior of the plant site on a regular basis provide the ultimate in site area security.

The next layer of physical security is the outside wall of the target structure itself. Notwithstanding door, window, and/or skylight entries, walls prevent most intruders from easy entry. If doors can only be entered using card-reader access, security is shored up or enhanced to an extent. The same can be said for windows and skylights that are fashioned small enough to prohibit normal human entry. These same "weak" spots in buildings can be bastioned with break-proof or reinforced security glass.

The final layer of security is provided by properly designed interior features of buildings. Examples of these types of features include internal doors and walls, equipment cages, and backup or redundant equipment.

In the preceding discussion, the conditions described refer to perfect-world conditions; that is, to those conditions that we "would want" (i.e., the security manager's proverbial wish list) to be incorporated into the design and installation of new ESS infrastructure. Post-9/11, in a not-so-perfect world, however, many of the peripheral (fence-line) measures described above are more difficult to incorporate into ESS infrastructure. This is not to say that ESS sites and facilities do not have fence lines or fences; many of them do. These fences are designed to keep vandals, thieves, and trespassers out. One problem, though, is the fact that many of these facilities were constructed several years ago, before urban encroachment literally encircled many of these sites, allowing, at present, little room for security stand-backs to be incorporated into electrical power stations, plants, and critical equipment locations. Based on personal observation, many of these fences face busy city streets or closely abut structures outside the fence line. The point is that when one sits down to plan a security upgrade, these factors must be taken into account.

Managers of ESS infrastructure have four primary security areas to manage. These security areas are listed and described below.

- Physical Security: In the ESS, physical security techniques and practices have the most effect at "fenced" locations. At such locations, a systems approach is best, where detection, assessment, communication, and response are planned and supported by resources, procedures, and policies.
- Cyber/Information Technology Security: Only the use of SCADA and other key operating systems that have been properly vetted and scrubbed of alleged Chinese and/or Russian Trojan horses hacked into the North American electrical grid is important. The only positive way to ensure the security of the North American grid is to disconnect its cyber and other digital systems from the internet. This step is impractical at the present time, but it points to the need to conduct frequent audits of the system and install firewall protection in SCADA and other systems to prevent hacking. Frequent third-party penetration testing is advised.
- Employment Screening: This mitigates the threat from the enemy at the water cooler (inside the organization). We are always amazed whenever we conduct security audits for various companies. Often a simple check, such as reviewing an employee's driving record, reveals that an employee

has no license, is driving on a suspended license, or has a horrific driving record. Hiring standards and pre-employment background investigations may help ensure the trustworthiness and reliability of personnel who have unescorted access to critical facilities.

- Protecting Potentially Sensitive Information: Benjamin Franklin wrote, "Three may keep a secret, if two of them are dead." This makes the point that reducing the likelihood that information could be used by those intent on disrupting operations or causing death and destruction at ESS sites is crucial. Information should only be shared within an organization on a need-to-know basis.

For existing facilities, security upgrades should be based on the results generated from a vulnerability assessment, which characterizes and prioritizes those assets that may be targeted. The vulnerabilities identified must be protected.

In the following sections, various security hardware and devices are described. Keep in mind that not all of the following security devices are employed everywhere; instead, facilities use some of those described as needed. The devices serve the main purpose of providing security against physical or digital intrusion. That is, they are designed to delay and deny intrusion and are normally coupled with detection and assessment technology. Possible additional security measures, based on the vulnerability assessment that may be recommended (covered in this text), include the following (NAERC, 2002):

- Electronic security
- Closing nonessential perimeter and internal portals
- Physical barriers such as bollards or Jersey walls
- Fencing
- Lighting
- Security surveys
- Vulnerability assessments
- Availability of security resources
- General personnel and security officer training
- Law-enforcement liaison
- Ensuring the availability of essential spare parts (machines, repair parts, wire, pipe, valves, transformers, etc.) for critical facilities

Keep in mind, however, and as mentioned previously, that no matter the type of security device or system employed, ESS systems cannot be made immune to all possible intrusions or attacks. Whenever a plant or facility safety/

security manager tells us that he or she has secured their site 100 percent, we are reminded of Schneier's (2000) view of security: "You can't defend. You can't prevent. The only thing you can do is detect and respond." Simply put, when it comes to making "anything" absolutely secure from intrusion or attack, there is inherently, or otherwise, no silver bullet.

In the next section, security hardware devices are discussed in detail. Before we describe these hardware devices, keep in mind that in addition to security hardware devices to help protect and monitor ESS assets, there are a few employee practices and actions that can be taken to protect emergency services assets. For example, when an organization's computer system fails and must be disposed of, how is it disposed of? Is there a procedure or practice in place to prevent the valuable information on the system's hard drive from being pulled from a trash heap or from a dumpster dive and used by potential enemies? Are shredders used? Are they state-of-the-art shredders that prevent scraps from being reassembled by enemies? Are building cleaning crews properly vetted and supervised? Do you have a team that routinely inspects suspended ceilings for bugs, cameras, and listening recorders? Do you have keystroke reader capability; that is, can you record what messages are being sent by employees? Do you routinely check hard-wired phone lines? Have you removed all door signs that tell anyone what is on the other side of the door? Have you trained your employees to be slightly suspicious of just about anything and everything? Have you opened your manholes lately to see what is inside?

All of these practices just mentioned do not require security hardware such as barriers, motion detectors, fences, locks, biometric systems, video cameras, armed guards, electrified fences, and so forth. What they require instead is common sense, awareness, and alert and engaged supervisors and employees. The point is that what is heard at the water cooler is sometimes more significant than anything a security alarm apparatus can provide.

SECURITY HARDWARE DEVICES

USEPA (2005) groups the infrastructure security devices and products described below into four general categories:[1]

- Physical asset monitoring and control devices
- Communication/integration
- Cyber protection devices
- Environmental monitoring devices

Physical Asset Monitoring and Control Devices

Aboveground, Outdoor Equipment Enclosures

ESS facilities and sites can consist of multiple structural components spread over a wide area and typically include a centralized production and distribution center, as well as component storage facilities that are typically distributed at multiple locations throughout the site. One of the primary reasons for constructing structural components that house operational equipment aboveground is that it eliminates the safety risks associated with confined-space entry, which is often required for the maintenance of equipment located belowground. In addition, space restrictions often limit the amount of equipment that can be located inside, and there are concerns that some types of equipment (such as backflow prevention devices—to prevent chemicals and fuel wastes from entering plant and off-site potable water systems) can, under certain circumstances, discharge fuel slurry or waste mixtures that could flood pits, vaults, or equipment rooms. With regard to electrical power, electrical substations are not usually suited for underground installation. Therefore, many pieces of critical electrical operational equipment are located outdoors and aboveground in configurations that are properly fenced, insulated, and isolated to prevent accidental electrical shock or short circuits/fires in equipment.

Experience demonstrates that many different system components can be and often are installed outdoors and aboveground, many of them controlled by wireless communication devices. Examples of these types of components include:

- Backflow prevention devices
- Air release and control valves
- Pressure vacuum breakers
- Oil and gas pumps and motors
- Petrochemical storage and feed equipment
- Meters
- Sampling equipment
- Instrumentation
- Electrical substations
- Oil and natural gas pipelines

One of the most effective security measures for protecting aboveground equipment, where feasible, is to place it inside a building or exterior fenced structure. Where this is not possible, enclosing the equipment or parts of the equipment using some sort of commercial or homemade add-on structure may help to prevent tampering with the equipment. These types of add-on structures or enclosures, which are designed to protect people and animals from electrocution and to protect equipment both from the elements and from unauthorized access or tampering, typically consist of a boxlike fenced structure that is placed over or around the entire component, or over/around critical parts of the component (i.e., valves, etc.), and is then secured to delay or prevent intruders from tampering with the equipment. The enclosures are typically locked or otherwise anchored to a solid foundation, which makes it difficult for unauthorized personnel to remove the enclosure and access the equipment.

Standardized aboveground enclosures are available in a wide variety of materials, sizes, and configurations. Many options and security features are also available for each type of enclosure, and this allows system operators the flexibility to customize an enclosure for a specific application and/or price range. In addition, most manufacturers can custom-design enclosures if standard, off-the-shelf enclosures do not meet a user's needs.

Many of these enclosures are designed to meet certain standards. For example, the American Society of Sanitary Engineers (ASSE) has developed Standard 1060, "Performance Requirements for Outdoor Enclosures for Backflow Prevention Assemblies." If an enclosure will be used to house a backflow preventer, this standard specifies the acceptable construction materials for the enclosure, as well as the performance requirements that the enclosure should meet, including specifications for freeze protection, drainage, air inlets, access for maintenance, and hinge requirements. ASSE 1060 also states that the enclosure should be lockable to enhance security.

Electrical substation and electrical equipment enclosures must meet the requirements and recommendations of the Occupational Safety and Health Administration (OSHA), the National Fire Protection Association (NFPA), the National Electrical Codes (NEC), the Institute of Electrical and Electronic Engineers (IEEE), and local code requirements.

Equipment enclosures can generally be categorized into one of four main configurations:

- One-piece, drop-over enclosures
- Hinged or removable top enclosures
- Sectional enclosures
- Shelters with access locks

All enclosures, including those with integral floors, must be secured to a foundation to prevent them from being moved or removed. Un- or poorly anchored enclosures may be blown off the equipment being protected or may be defeated by intruders. In either case, this may result in the equipment beneath the enclosure becoming exposed and damaged. Therefore, ensuring that the enclosure is securely anchored will increase the security of the protected equipment.

The three basic types of foundation that can be used to anchor the aboveground equipment enclosure are concrete footers, concrete slabs-on-grade, or manufactured fiberglass pads. The most common types of foundation utilized for equipment enclosures are standard or slab-on-grade footers; however, local climate and soil conditions may dictate whether either of these types of foundation can be used. These foundations can be either precast or poured in place at the installation site. Once the foundation is installed and properly cured, the equipment enclosure is bolted or anchored to the foundation to secure it in place.

An alternative foundation, specifically for use with smaller Hot Box enclosures, is a manufactured fiberglass pad known as the Glass Pad. The Glass Pad has the center cut out so that it can be dropped directly over the piece of equipment being enclosed. Once the pad is set level on the ground, it is backfilled over a two-inch flange located around its base. The enclosure is then placed on top of the foundation and is locked in place with either a staple anchor or a slotted anchor, depending on the enclosure configuration.

One of the primary attributes of a security enclosure is its strength and resistance to breaking and penetration. Accordingly, the materials from which the enclosure is constructed will be important in determining the strength of the enclosure, and thus its usefulness for security applications. Enclosures are typically manufactured of either fiberglass or aluminum. With the exception of the one-piece, drop-over enclosure, which is typically fabricated from fiberglass, each configuration described above can be constructed from either material. In addition, enclosures can be custom manufactured from polyurethane, galvanized steel, or stainless steel. Galvanized or stainless steel is often offered as an exterior layer, or "skin," for an aluminum enclosure. Although they are typically utilized in underground applications, precast concrete structures can also be used as aboveground equipment enclosures. However, precast structures are much heavier and more difficult to maneuver than are their fiberglass and aluminum counterparts. Concrete is also brittle, and that can be a security concern; however, products such as epoxy coating can be applied to concrete structures to add strength and minimize security risks. Because precast concrete structures can be purchased from any concrete producer, this document does not identify specific vendors for these types of products.

In addition to the construction materials, enclosure walls can be configured or reinforced to give them added strength. Adding insulation is one option that can strengthen the structural characteristics of an enclosure; however, some manufacturers offer additional features to add strength to exterior walls. For example, while most enclosures are fabricated with a flat wall construction, some vendors manufacture fiberglass shelters with ribbed exterior walls. These ribs increase the structural integrity of the wall and allow the fabrication of standard shelters up to twenty feet in length. Another vendor has developed a proprietary process that uses a series of integrated fiberglass beams that are placed throughout a foam inner core to tie together the interior and exterior walls and roof. Yet another vendor constructs aluminum enclosures with horizontal and vertical redwood beams for structural support.

Other security features that can be implemented on aboveground, outdoor equipment enclosures include locks, mounting brackets, tamper-resistant doors, and exterior lighting.

Active Security Barriers (Crash Barriers)

Active security barriers (also known as crash barriers) are large structures that are placed in roadways at entrance and exit points of protected facilities to control vehicle access to these areas. These barriers are placed perpendicular to traffic so that the only way that traffic can pass the barrier is for the barrier to be moved out of the roadway. These types of barriers are typically constructed from sturdy materials, such as concrete or steel, such that vehicles cannot penetrate through them. They are also designed at a certain height off the roadway so that vehicles cannot go over them.

The key difference between active security barriers (e.g., wedges, crash beams, gates, retractable bollards, portable barricades) and passive security barriers (e.g., nonmovable bollards, Jersey barriers, planters) is that active security barriers are designed so that they can be raised and lowered or moved out of the roadway easily to allow authorized vehicles to pass them. Many of these types of barriers are designed so that they can be opened and closed automatically (e.g., mechanized gates, hydraulic wedge barriers), while others are easy to open and close manually (e.g., swing crash beams, manual gates). In contrast to active barriers, passive barriers are permanent, nonmovable barriers, and thus they are typically used to protect the perimeter of a protected facility, such as sidewalks and other areas that do not require vehicular traffic to pass them. Several of the major types of active security barriers such as wedge barriers, crash beams, gates, bollards, and portable/removable barricades are described below.

Wedge barriers are plated, rectangular steel buttresses approximately two to three feet high that can be raised and lowered from the roadway.

When they are in the open position, they are flush with the roadway, and vehicles can pass over them. However, when they are in the closed (armed) position, they project up from the road at a forty-five-degree angle, with the upper end pointing toward the oncoming vehicle and the base of the barrier facing away from the vehicle. Generally, wedge barriers are constructed from heavy-gauge steel or from concrete that contains an impact-dampening iron rebar core that is strong and resistant to breaking or cracking, thereby allowing them to withstand the impact of a vehicle attempting to crash through them. In addition, both of these materials help to transfer the energy of the impact over the barrier's entire volume, thus helping to prevent the barrier from being sheared off its base. In addition, because the barrier is angled away from traffic, the force of any vehicle impacting the barrier is distributed over the entire surface of the barrier and is not concentrated at the base, which helps prevent the barrier from breaking off at the base. Finally, the angle of the barrier helps to hang up any vehicle attempting to drive over it.

Wedge barriers can be fixed or portable. Fixed wedge barriers can be mounted on the surface of the roadway (surface-mounted wedges) or in a shallow mount in the road's surface, or they can be installed completely below the road surface. Surface-mounted wedge barricades operate by rising from a flat position on the surface of the roadway, while shallow-mount wedge barriers rise from their resting position just below the road surface. In contrast, below-surface wedge barriers operate by rising from beneath the road surface. Both the shallow-mounted and surface-mounted barriers require little or no excavation and thus do not interfere with buried utilities. All three barrier mounting types project above the road surface and block traffic when they are raised into the armed position. Once they are disarmed and lowered, they are flush with the road, thereby allowing traffic to pass. Portable wedge barriers are moved into place on wheels that are removed after the barrier has been set in place.

Installing rising wedge barriers requires preparation of the road surface. Installing surface-mounted wedges does not require that the road be excavated; however, the road surface must be intact and strong enough to allow the bolts anchoring the wedge to the road surface to attach properly. Shallow-mount and below-surface wedge barricades require excavation of a pit that is large enough to accommodate the wedge structure, as well as any arming/disarming mechanism. Generally, the bottom of the excavation pit is lined with gravel to allow for drainage. Areas not sheltered from rain or surface runoff can install a gravity drain or self-priming pump. Table 8.1 lists the pros and cons of wedge barriers.

Table 8.1. Pros and Cons of Wedge Barriers

Pros	*Cons*
Can be surface mounted or completely installed below the roadway surface.	Installations below the surface of the roadway will require construction that may interfere with buried utilities.
Wedge barriers have a quick response time (normally 3.5–10.5 seconds, but the response time can be 1–3 seconds in emergency situations). Because emergency activation of the barrier causes more wear and tear on the system than does normal activation, it is recommended for use only in true emergency situations.	Regular maintenance is needed to keep the wedge fully operational.
Surface or shallow-mount wedge barricades can be utilized in locations with a high water table and/or corrosive soils.	Improper use of the system may result in authorized vehicles being hung up by the barrier and damaged. Guards must be trained to use the system properly to ensure that this does not happen. Safety technologies may also be installed to reduce the risk of the wedge activating under an authorized vehicle.
All three wedge barrier designs have a high crash rating, thereby allowing them to be employed for higher-security applications.	
These types of barrier are extremely visible, which may deter potential intruders.	

Source: USEPA (2005).

Crash beam barriers consist of aluminum beams that can be opened or closed across the roadway. While there are several different crash beam designs, every crash beam system consists of an aluminum beam that is supported on each side by a solid footing or buttress, which is typically constructed from concrete, steel, or some other strong material. Beams typically contain an interior steel cable (typically at least one inch in diameter) to give the beam added strength and rigidity. The beam is connected by a heavy-duty hinge or other mechanism to one of the footings so that it can swing or rotate out of the roadway when it is open and can swing back across the road when it is in the closed (armed) position, blocking the road and inhibiting access by unauthorized vehicles. The non-hinged end of the beam can be locked into its footing, thus providing anchoring for the beam on both sides of the road and increasing the beam's resistance to any vehicles attempting to penetrate through it. In addition, if the crash beam is hit by a vehicle, the aluminum beam transfers the impact energy to the interior cable, which in turn transfers the impact energy through the footings and into their foundation, thereby minimizing the chance that the impact will snap the beam and allow the intruding vehicle to pass through.

Crash beam barriers can employ drop-arm, cantilever, or swing-beam designs. Drop-arm crash beams operate by raising and lowering the beam vertically across the road. Cantilever crash beams are projecting structures that are opened and closed by extending the beam from the hinge buttress to the receiving buttress located on the opposite side of the road. In the swing-beam design, the beam is hinged to the buttress such that it swings horizontally across the road. Generally, swing-beam and cantilever designs are used at locations where a vertical lift beam is impractical. For example, the swing beam or cantilever designs are utilized at entrances and exits with overhangs, trees, or buildings that would physically block the operation of the drop-arm beam design.

Installing any of these crash beam barriers involves the excavation of a pit approximately forty-eight inches deep for both the hinge and the receiver footings. Due to the depth of excavation, the site should be inspected for underground utilities before digging begins. Table 8.2 lists the pros and cons of crash beams.

Table 8.2. Pros and Cons of Crash Beams

Pros	*Cons*
Requires little maintenance, while providing long-term durability.	Crash beams have a slower response time (normally 9.5–15.3 seconds, but can be reduced to 7–10 seconds in emergency situations) than do other types of active security barriers, such as wedge barriers. Because emergency activation of the barrier causes more wear and tear on the system than does normal activation, it is recommended for use only in true emergency situations.
No excavation is required in the roadway itself to install crash beams.	All three crash beam designs possess a low crash rating relative to other types of barriers, such as wedge barriers, and thus they typically are used for lower-security applications.
	Certain crash barriers may not be visible to oncoming traffic and therefore may require additional lighting and/or other warning markings to reduce the potential for traffic to accidentally run into the beam.

Source: USEPA (2005).

In contrast to wedge barriers and crash beams, which are typically installed separately from a fence line, *gates* are often integrated units of a perimeter fence or wall around a facility. Gates are basically movable pieces of fencing that can be opened and closed across a road. When the gate is in the closed (armed) position, the leaves of the gate lock into steel buttresses that are embedded in concrete foundations located on both sides of the roadway, thereby blocking access to the roadway. Generally, gate barricades are constructed from a combination of heavy-gauge steel and aluminum that can absorb an impact from vehicles attempting to ram through them. Any remaining impact energy not absorbed by the gate material is transferred to the steel buttresses and their concrete foundation.

Gates can utilize a cantilever, linear, or swing design. Cantilever gates are projecting structures that operate by extending the gate from the hinge footing across the roadway to the receiver footing. A linear gate is designed to slide across the road on tracks via a rack-and-pinion drive mechanism. Swing gates are hinged so that they can swing horizontally across the road.

Installation of the cantilever, linear, or swing gate designs described above involves the excavation of a pit approximately forty-eight inches deep for both the hinge and receiver footings to which the gates are attached. Due to the depth of excavation, the site should be inspected for underground utilities before digging begins. Table 8.3 lists the pros and cons of gates.

Table 8.3. Pros and Cons of Gates

Pros	*Cons*
All three gate designs possess an intermediate crash rating, thereby allowing them to be utilized for medium- to higher-security applications.	Gates have a slower response time (normally 10–15 seconds, but can be reduced to 7–10 seconds in emergency situations) than do other types of active security barriers, such as wedge barriers. Because emergency activation of the barrier causes more wear and tear on the system than does normal activation, it is recommended for use only in true emergency situations.
Requires very little maintenance.	
Can be tailored to blend in with perimeter fencing.	
Gate construction requires no roadway excavation.	
Cantilever gates are useful for roads with high crowns or drainage gutters.	
These types of barriers are extremely visible, which may deter intruders.	
Gates can also be used to control pedestrian traffic.	

Source: USEPA (2005).

Bollards are vertical barriers at least three feet tall and 0.4 to 2 feet in diameter that are typically set four to five feet apart from each other so that they block vehicles from passing between them. Smaller bollards, usually four-inch-diameter pipes filled with concrete, are installed in parking areas to prevent vehicles from striking walls or windows or to protect walkway areas. Bollards can either be fixed in place, removable, or retractable. Fixed and removable bollards are passive barriers that are typically used along building perimeters or on sidewalks to keep vehicles off while allowing pedestrians to pass. In contrast to passive bollards, retractable bollards are active security barriers that can easily be raised and lowered to allow vehicles to pass over them. Thus, they can be used in driveways or on roads to control vehicular access. When the bollards are raised, they project above the road surface and block the roadway; when they are lowered, they sit flush with the road surface and thus allow traffic to pass over them. Retractable bollards are typically constructed from steel or other materials that have a low weight-to-volume ratio so that they require low power to raise and lower them. Steel is also more resistant to breaking than is a more brittle material, such as concrete, and is better able to withstand direct vehicular impact without breaking apart.

Retractable bollards are installed in a trench dug across a roadway—typically at an entrance or gate. Installing retractable bollards requires preparing the road surface. Depending on the vendor, bollards can be installed either in a continuous slab of concrete or in individual excavations with concrete poured in place. The required excavation for a bollard is typically slightly wider than the bollard width and slightly deeper than the bollard height when extended aboveground. The bottom of the excavation is typically lined with gravel to allow drainage. The bollards are then connected to a control panel that controls the raising and lowering of the bollards. Installation typically requires mechanical, electrical, and concrete work; if utility personnel with these skills are available, then the utility can install the bollards themselves. Table 8.4 lists the pros and cons of retractable bollards.

Table 8.4. Pros and Cons of Retractable Bollards

Pros	*Cons*
Bollards have a quick response time (normally 3–10 seconds, but can be reduced to 1–3 seconds in emergency situations).	Bollard installations will require construction below the surface of the roadway, which may interfere with buried utilities.
Bollards have an intermediate crash rating, which allows them to be utilized for medium- to higher-security applications.	Some maintenance is needed to ensure that the barrier is free to move up and down.
	The distance between bollards must be decreased (i.e., more bollards must be installed along the same perimeter) to make these systems effective against small vehicles (e.g., motorcycles).

Source: USEPA (2005).

Portable/removable barriers, which can include removable crash beams and wedge barriers, are mobile obstacles that can be moved in and out of position on a roadway. For example, a crash beam may be completely removed and stored off-site when it is not needed. An additional example would be wedge barriers that are equipped with wheels that can be removed after the barricade is towed into place.

When portable barricades are needed, they can be moved into position rapidly. To provide them with added strength and stability, they are typically anchored to buttress boxes that are located on either side of the road. These buttress boxes, which may or may not be permanent, are usually filled with sand, water, cement, or gravel to make them heavy and aid in stabilizing the portable barrier. In addition, these buttresses can help dissipate any impact energy from vehicles crashing into the barrier itself.

Because these barriers are not anchored into the roadway, they do not require excavation or other related construction for installation. In contrast, they can be assembled and made operational in a short period of time. The primary shortcoming of this type of design is that these barriers may move if they are hit by vehicles. Therefore, it is important to carefully assess the placement and anchoring of these types of barriers to ensure that they can withstand the types of impacts that may be anticipated at that location. Table 8.5 lists the pros and cons of portable/removable barricades.

Because the primary threat to active security barriers is that vehicles will attempt to crash through them, their most important attributes are their size, strength, and crash resistance. Other important features for an active security

Table 8.5. Pros and Cons of Portable/Removable Barricades

Pros	*Cons*
Installing portable barricades requires no foundation or roadway excavation.	Portable barriers may move slightly when hit by a vehicle, resulting in a lower crash resistance.
Can be moved in and out of position in a short period of time.	Portable barricades typically require 7.75–16.25 seconds to move into place, and thus they are considered to have a medium response time when compared with other active barriers.
Wedge barriers equipped with wheels can be easily towed into place.	
Minimal maintenance is needed to keep barriers fully operational.	

Source: USEPA (2005).

barrier are the mechanisms by which the barrier is raised and lowered to allow authorized vehicle entry, as well as weather resistance and safety features.

Alarms

An *alarm system* is a type of electronic monitoring system that is used to detect and respond to specific types of events—such as unauthorized access to an asset or a possible fire. In chemical processing systems, alarms are also used to alert operators when process operating or monitoring conditions go out of preset parameters (i.e., process alarms). These types of alarms are primarily integrated with process monitoring and reporting systems (i.e., SCADA systems). Note that this discussion does not focus on alarm systems that are not related to a facility's processes.

Alarm systems can be integrated with fire detection systems, intrusion detection systems (IDSs), access control systems, or closed-circuit television (CCTV) systems, such that these systems automatically respond when the alarm is triggered. For example, a smoke detector alarm can be set up to automatically notify the fire department when smoke is detected, or an intrusion alarm can automatically trigger cameras to turn on in a remote location so that personnel can monitor that location.

An alarm system consists of sensors that detect different types of events; an arming station that is used to turn the system on and off; a control panel that receives information, processes it, and transmits the alarm; and an annunciator that generates a visual and/or audible response to the alarm. When a sensor is tripped, it sends a signal to a control panel, which triggers a visual or audible alarm and/or notifies a central monitoring station. A more complete description of each of the components of an alarm system is provided below.

Detection devices (also called *sensors*) are designed to detect a specific type of event (such as smoke, intrusion, etc.). Depending on the type of event they are designed to detect, sensors can be located inside or outside the facility or other asset. When an event is detected, the sensors use some type of communication method (such as wireless radio transmitters, conductors, or cables) to send signals to the control panel to generate the alarm. For example, a smoke detector sends a signal to a control panel when it detects smoke.

Alarms use either normally closed (NC) or normally open (NO) electric loops, or "circuits," to generate alarm signals. These two types of circuit are discussed separately below.

In NC loops or circuits, all of the system's sensors and switches are connected in series. The contacts are "at rest" in the closed (on) position, and current continually passes through the system. However, when an event triggers the sensor, the loop is opened, breaking the flow of current through the system and triggering the alarm. NC switches are used more often than NO switches because the alarm will be activated if the loop or circuit is broken or cut, thereby reducing the potential for circumventing the alarm. This is known as a "supervised" system.

In NO loops or circuits, all of the system's sensors and switches are connected in parallel. The contacts are "at rest" in the open (off) position, and no current passes through the system. However, when an event triggers the sensor, the loop is closed. This allows current to flow through the loop, triggering the alarm. NO systems are not "supervised" because the alarm will not be activated if the loop or circuit is broken or cut. However, adding an end-of-line resistor to an NO loop will cause the system to alarm if tampering is detected.

An *arming station*, which is the main user interface with the security system, allows the user to arm (turn on), disarm (turn off), and communicate with the system. How a specific system is armed will depend on how it is used. For example, while IDSs can be armed for continuous operation (twenty-four hours a day), they are usually armed and disarmed according to the work schedule at a specific location so that personnel going about their daily activities do not set off the alarms. In contrast, fire protection systems are typically armed twenty-four hours a day.

A *control panel* receives information from the sensors and sends it to an appropriate location, such as to a central operations station or to a twenty-four-hour monitoring facility. Once the alarm signal is received at the central monitoring location, personnel monitoring for alarms can respond (such as by sending security teams to investigate or by dispatching the fire department).

An *annunciator* responds to the detection of an event by emitting a signal. This signal may be visual, auditory, electronic, or a combination of these. For example, fire alarm signals will always be connected to audible annunciators, whereas intrusion alarms may not be.

Alarms can be reported locally, remotely, or both. Local and remotely (centrally) reported alarms are discussed in more detail below.

A *local alarm* emits a signal at the location of the event (typically using a bell or siren). A "local only" alarm emits a signal at the location of the event but does not transmit the alarm signal to any other location (i.e., it does not transmit the alarm to a central monitoring location). Typically, the purpose of a "local only" alarm is to frighten away intruders, and possibly to attract the attention of someone who might notify the proper authorities. Because no signal is sent to a central monitoring location, personnel can only respond to a local alarm if they are in the area and can hear and/or see the alarm signal.

Fire alarm systems must have local alarms, including both audible and visual signals. Most fire alarm signal and response requirements are codified in the National Fire Alarm Code, National Fire Protection Association (NFPA) 72. NFPA 72 discusses the application, installation, performance, and maintenance of protective signaling systems and their components. In contrast to fire alarms, which require a local signal when fire is detected, many IDSs do not have a local alert device, because monitoring personnel do not wish to inform potential intruders that they have been detected. Instead, these types of systems silently alert monitoring personnel that an intrusion has been detected, thus allowing monitoring personnel to respond.

In contrast to systems that are set up to transmit "local only" alarms when the sensors are triggered, systems can also be set up to transmit signals to a *central location,* such as to a control room or guard post at the utility, or to a police or fire station. Most fire/smoke alarms are set up to signal both at the location of the event and at a fire station or central monitoring station. Many insurance companies require that facilities install certified systems that include alarm communication to a central station. For example, systems certified by the Underwriters Laboratories (UL) require that the alarm be reported to a central monitoring station.

The main difference between alarm systems lies in the types of event-detection devices used in different systems. *Intrusion sensors*, for example, consist of two main categories: perimeter sensors and interior (space) sensors. *Perimeter intrusion sensors* are typically applied on fences, doors, walls, windows, etc., and are designed to detect an intruder before he/she accesses a protected asset (i.e., perimeter intrusion sensors are used to detect intruders attempting to enter through a door, window, etc.). In contrast, *interior intrusion sensors* are designed to detect an intruder who has already accessed the protected asset (i.e., interior intrusion sensors are used to detect intruders once they are already within a protected room or building). These two types of detection devices can be complementary, and they are often used together to enhance security for an asset. For example, a typical intrusion alarm system might employ a perimeter glass-break detector that protects against intruders accessing a room through a window, as well as an ultrasonic interior sensor that detects intruders that have gotten into the room without using the window. Table 8.6 lists and describes types of perimeter and interior sensors.

Table 8.6. Perimeter and Interior Sensors

Perimeter Sensor	*Description*
Foil	Foil is a thin, fragile, lead-based metallic tape that is applied to glass windows and doors. The tape is applied to the window or door, and electric wiring connects this tape to a control panel. The tape functions as a conductor and completes the electric circuit with the control panel. When an intruder breaks the door or window, the fragile foil breaks, opening the circuit and triggering an alarm condition.
Magnetic switches (reed switches)	The most widely used perimeter sensor. They are typically used to protect doors, as well as windows that can be opened (windows that cannot be opened are more typically protected by foil alarms).
Glass-break detectors	Placed on glass and senses vibrations in the glass when it is disturbed. The two most common types of glass-break detectors are shock sensors and audio discriminators.
Interior Sensor	*Description*
Passive infrared (PIR)	Presently the most popular and cost-effective interior sensors. PIR detectors monitor infrared radiation (energy in the form of heat) and detect rapid changes in temperature within a protected area. Because infrared radiation is emitted by all living things, these types of sensors can be very effective.
Quad PIRs	Consist of two dual-element sensors combined in one housing. Each sensor has a separate lens and a separate processing circuitry, which allows each lens to be set up to generate a different protection pattern.
Ultrasonic detectors	Emit high-frequency sound waves and sense movement in a protected area by sensing changes in these waves. The sensor emits sound waves that stabilize and set a baseline condition in the area to be protected. Any subsequent movement within the protected area by a would-be intruder will cause a change in these waves, thus creating an alarm condition.
Microwave detectors	Emit ultrahigh-frequency radio waves, and the detector senses any changes in these waves as they are reflected throughout the protected space. Microwaves can penetrate through walls, and thus a unit placed in one location may be able to protect multiple rooms.
Dual technology devices	Incorporate two different types of sensor technology (such as PIR and microwave technology) together in one housing. When both technologies sense an intrusion, an alarm is triggered.

Source: USEPA (2005).

Fire detection/fire alarm systems consist of different types of fire detection and fire alarm devices available. These systems may detect fire, heat, smoke, or a combination of any of these. For example, a typical fire alarm system might consist of heat sensors, which are located throughout a facility and which detect high temperatures or a certain change in temperature over a fixed time period. A different system might be outfitted with both smoke and heat detection devices. A summary of several different types of fire/smoke/heat detection sensors is provided in table 8.7.

Table 8.7. Fire/Smoke/Heat Detection Sensors

Detector Type	*Description*
Thermal detector	Sense when temperatures exceed a set threshold (fixed temperature detectors) or when the rate of change of temperature increases over a fixed time period (rate-of-rise detectors).
Duct detector	Is located within the heating and ventilation ducts of the facility. This sensor detects the presence of smoke within the system's return or supply ducts. A sampling tube can be added to the detector to help span the width of the duct.
Smoke detectors	Sense invisible and/or visible products of combustion. The two principle types of smoke detectors are photoelectric and ionization detectors. The major differences between these devices are described below: • Photoelectric smoke detectors react to visible particles of smoke. These detectors are more sensitive to the cooler smoke with large smoke particles that is typical of smoldering fires. • Ionization smoke detectors are sensitive to the presence of ions produced by the chemical reactions that take place with few smoke particles, such as those typically produced by fast-burning/flaming fires.
Carbon monoxide (CO) detectors	Are used to indicate the outbreak of fire by sensing the level of carbon monoxide in the air. The detector has an electrochemical cell that senses carbon monoxide, but not some of the other products of combustion.
Beam detectors	Are designed to protect large, open spaces such as industrial warehouses. These detectors consist of three parts: the transmitter, which projects a beam of infrared light; the receiver, which registers the light and produces an electrical signal; and the interface, which processes the signal and generates alarm or fault signals. In the event of a fire, smoke particles obstruct the beam of light. Once a preset threshold is exceeded, the detector will go into alarm.
Flame detectors	Sense either ultraviolet (UV) or infrared (IR) radiation emitted by a fire.
Air-sampling detectors	Actively and continuously sample the air from a protected space and are able to sense the precombustion stages of incipient fire.

Source: USEPA (2005).

Once a sensor in an alarm system detects an event, it must communicate an alarm signal. The two basic types of alarm communication systems are hardwired and wireless. Hardwired systems rely on wire that is run from the control panel to each of the detection devices and annunciators. Wireless systems transmit signals from a transmitter to a receiver through the air—primarily using radio or other waves. Hardwired systems are usually lower cost, more reliable (they are not affected by terrain or environmental factors), and significantly easier to troubleshoot than wireless systems. However, a major disadvantage of hardwired systems is that it may not be possible to hardwire all locations (for example, it may be difficult to hardwire remote locations). In addition, running wires to the required locations can be both time consuming and costly. The major advantage to using wireless systems is that they can often be installed in areas where hardwired systems are not feasible. However, wireless components can be much more expensive when compared to hardwired systems. In addition, in the past, it has been difficult to perform self-diagnostics on wireless systems to confirm that they are communicating properly with the controller. Presently, the majority of wireless systems incorporate supervising circuitry, which allows the subscriber to know immediately if there is a problem with the system (such as a broken detection device or a low battery), or if a protected door or window has been left open.

Backflow Prevention Devices

Backflow prevention devices are designed to prevent backflow, which is the reversal of the normal and intended direction of water flow in a water system. Backflow is a potential problem in a petrochemical processing system because if incorrectly cross-connected to potable water, it can spread contaminated water back through a distribution system. For example, backflow at uncontrolled cross-connections (cross-connections are any actual or potential connection between the public water supply and a source of chemical contamination) can allow pollutants or contaminants to enter the potable water system. More specifically, backflow from private plumbing systems, industrial areas, hospitals, and other hazardous contaminant–containing systems into public water mains and wells poses serious public health risks and security problems. Cross-contamination from private plumbing systems can contain biological hazards (such as bacteria or viruses) or toxic substances that can contaminate and sicken an entire population in the event of backflow. The majority of historical incidences of backflow have been accidental, but growing concern that contaminants could be intentionally backfed into a system is prompting increased awareness for private homes, businesses, industries, and areas most vulnerable to intentional strikes. Therefore, backflow prevention is a major tool for the protection of water systems.

Backflow may occur under two types of conditions: backpressure and backsiphonage. *Backpressure* is the reverse of normal flow direction within a piping system that is the result of the downstream pressure being higher than the supply pressure. Such reductions in the supply pressure occur whenever the amount of water being used exceeds the amount of water supplied, such as during water-line flushing, firefighting, or breaks in water mains. *Backsiphonage* is the reverse of normal flow direction within a piping system that is caused by negative pressure in the supply piping (i.e., a vacuum or partial vacuum within the water supply piping). Backsiphonage can occur where there is a high velocity in a pipeline; when there is a line repair or break that is lower than a service point; or when there is lowered main pressure due to high water withdrawal rate, such as during firefighting or water-main flushing.

To prevent backflow, various types of backflow preventers are appropriate for use. The primary types of backflow preventers are:

- Air gap drains
- Double-check valves
- Reduced pressure principle assemblies
- Pressure vacuum breakers

Biometric Security Systems

Biometrics involves measuring the unique physical characteristics or traits of the human body. In ancient times, biometrics involved the judging of one's accent, body hair, or face to determine friend or foe. Presently, it is well known that any aspect of the body that is measurably different from person to person—for example, fingerprints or eye characteristics—can serve as a unique biometric identifier for that individual. Biometric systems recognizing fingerprints, palm shape, eyes, face, voice, or signature comprise the bulk of the current biometric systems that exist.

Biometric security systems use biometric technology combined with some type of locking mechanism to control access to specific assets. In order to access an asset controlled by a biometric security system, an individual's biometric trait must be matched with an existing profile stored in a database. If there is a match between the two, the locking mechanisms (which could be a physical lock, such as at a doorway; an electronic lock, such as at a computer terminal; or some other type of lock) is disengaged, and the individual is given access to the asset.

A biometric security system is typically comprised of the following components:

- A sensor, which measures/records a biometric characteristic or trait.
- A control panel, which serves as the connection point between various system components. The control panel communicates information back and forth between the sensor and the host computer and controls access to the asset by engaging or disengaging the system lock based on internal logic and information from the host computer.
- A host computer, which processes and stores the biometric trait in a database.
- Specialized software, which compares an individual image taken by the sensor with a stored profile or profiles.
- A locking mechanism that is controlled by the biometric system.
- A power source to power the system.

Biometric Hand and Finger Geometry Recognition

Hand and finger geometry recognition is the process of identifying an individual through the unique "geometry" (shape, thickness, length, width, etc.) of that individual's hand or fingers. Hand geometry recognition has been employed since the early 1980s and is among the most widely used biometric technologies for controlling access to important assets. It is easy to install and use and is appropriate for any location requiring use of highly accurate, nonintrusive, two-finger biometric security. For example, it is currently used in numerous workplaces, day care facilities, hospitals, universities, airports, refineries, and power plants.

A newer option in hand geometry recognition technology is finger geometry recognition (not to be confused with fingerprint recognition). Finger geometry recognition relies on the same scanning methods and technologies as does hand geometry recognition, but the scanner only scans two of the user's fingers, as opposed to his entire hand. Finger geometry recognition has been in commercial use since the mid-1990s and is mainly used in time and attendance applications (i.e., to track when individuals have entered and exited a location). To date, the only large-scale commercial use of two-finger geometry for controlling access is at Disney World, where season pass holders use the geometry of their index and middle finger to gain access to the facilities.

To use a hand or finger geometry unit, an individual presents his or her hand or fingers to the biometric unit for "scanning." The scanner consists of a charge-coupled device (CCD), which is essentially a high-resolution digital camera; a reflective platen on which the hand is placed; and a mirror or mirrors that help capture different angles of the hand or fingers. The camera "scans" individual geometric characteristics of the hand or fingers by taking multiple images while the user's hand rests on the reflective platen. The camera also captures "depth," or three-dimensional information, through light reflected from the mirrors and the reflective platen. This live image is then compared to a "template" that was previously established for that individual when they were "enrolled" in the system. If the live scan of the individual matches the stored template, the individual is "verified" and is given access to that asset. Typically verification takes about two seconds. In access control applications, the scanner is usually connected to some sort of electronic lock, which unlocks the door, turnstile, or other entry barrier when the user is verified. The user can then proceed through the entrance. In time and attendance applications, the time that an individual checks in and out of a location is stored for later use.

As discussed above, hand and finger geometry recognition systems can be used in several different types of applications, including access control and time and attendance tracking. While time and attendance tracking can be used for security, it is primarily used for operations and payroll purposes (i.e., clocking in and clocking out). In contrast, access control applications are more likely to be security related. Biometric systems are widely used for access control and can be used on various types of assets, including entryways, computers, vehicles, etc. However, because of their size, hand/finger recognition systems are primarily used in entryway access control applications.

Biometric Iris Recognition

The iris, which is the colored or pigmented area of the eye surrounded by the sclera (the white portion of the eye), is a muscular membrane that controls the amount of light entering the eye by contracting or expanding the pupil (the dark center of the eye). The dense, unique patterns of connective tissue in the human iris were first noted in 1936, but it was not until 1994, when algorithms for iris recognition were created and patented, that commercial applications using biometric iris recognition began to be used extensively. There are now two vendors producing iris recognition technology: both the original developer of these algorithms and a second company, which has developed and patented a different set of algorithms for iris recognition.

The iris is an ideal characteristic for identifying individuals because it is formed in utero, and its unique patterns stabilize around eight months after birth. No two irises are alike, neither an individual's right and left irises, nor the irises of identical twins. The iris is protected by the cornea (the clear covering over the eye), and therefore it is not subject to the aging or physical changes (and potential variation) that are common to some other biometric measures, such as the hand, fingerprints, and the face. Although some limited changes can occur naturally over time, these changes generally occur in the iris's melanin and therefore affect only the eye's color, not its unique patterns (in addition, because iris scanning uses only black-and-white images, color changes would not affect the scan anyway). Thus, barring specific injuries or certain rare surgeries directly affecting the iris, the iris's unique patterns remain relatively unchanged over an individual's lifetime.

Iris recognition systems employ a monochromatic or black-and-white video camera that uses both visible and near infrared light to take video of an individual's iris. Video is used rather than still photography as an extra security procedure. The video is used to confirm the normal continuous fluctuations of the pupil as the eye focuses, which ensures that the scan is of a living human being and not a photograph or some other attempted hoax. A high-resolution image of the iris is then captured or extracted from the video, using a device often referred to as a "frame grabber." The unique characteristics identified in this image are then converted into a numeric code, which is stored as a "template" for that user.

Card Identification/Access/Tracking Systems

A card-reader system is a type of electronic identification system that is used to identify a card and then perform an action associated with that card. Depending on the system, the card may identify where a person is or where they were at a certain time; or it may authorize another action, such as disengaging a lock. For example, a security guard may use his card at card readers located throughout a facility to indicate that he has checked a certain location at a certain time. The reader will store the information and/or send it to a central location, where it can be checked later to ensure that the guard has patrolled the area. Other card-reader systems can be associated with a lock so that the cardholder must have their card read and accepted by the reader before the lock disengages.

A complete card-reader system typically consists of the following components:

- Access cards that are carried by the user
- Card readers, which read the card signals and send the information to control units
- Control units, which control the response of the card reader to the card
- A power source

A "card" may be a typical card or another type of device, such as a key fob or wand. These cards store electronic information, which can range from a simple code (i.e., the alphanumeric code on a proximity card) to individualized personal data (i.e., biometric data on a smart card). The card reader reads the information stored on the card and sends it to the control unit, which determines the appropriate action to take when a card is presented. For example, in a card access system, the control unit compares the information on the card versus stored access authorization information to determine if the card holder is authorized to proceed through the door. If the information stored in the card-reader system indicates that the key is authorized to allow entrance through the doorway, the system disengages the lock, and the key holder can proceed through the door.

There are many different types of card-reader systems on the market. The primary differences between card-reader systems are the way the data is encoded on the cards, the way the data is transferred between the card and the card reader, and the types of applications for which the systems are best suited. However, all card systems are similar in the way that the card reader and the control unit interact to respond to the card.

There are several types of technologies available for card-reader systems. These include:

- Proximity
- Wiegand
- Smart card
- Magnetic stripe
- Bar code
- Infrared
- Barium ferrite
- Hollerith
- Mixed technologies

Table 8.8 summarizes various aspects of card-reader technologies. The determination of the level of security rating (low, moderate, or high) is based on the level of technology of a given card-reader system and how simple it is to duplicate that technology and thus bypass the security. Vulnerability ratings are based on whether the card reader can be damaged easily due to frequent use or difficult working conditions (i.e., weather conditions if the reader is located outside). Often this is influenced by the number of moving parts in the system—the more moving parts, the greater the system's susceptibility to damage. The life-cycle rating is based on the durability of a given card-reader system over its entire operational period. Systems requiring frequent physical contact between the reader and the card often have a shorter life cycle due to the wear and tear to which the equipment is exposed. For many card-reader

Table 8.8. Card-Reader Technology

Type of Card Reader	*Technology*	*Life Cycle*	*Vulnerability*	*Level of Security*
Proximity	Embedded radio frequency circuits encoded with unique information	Long	Virtually none	Moderate–high
Wiegand	Short lengths of small-diameter, special-alloy wire with unique magnetic properties	Long	Low susceptibility to damage; high durability due to embedded wires	Moderate–expensive
Magnetic stripe	Electromagnetic charges to encode information on a piece of tape attached to back of card	Moderate	Moderately susceptible to damage due to frequency of use	Low–moderate
Barcode	Series of narrow and wide bars and spaces	Short	High; easily damaged	Low
Hollerith	Holes punched in a plastic or paper card and read optically	Short	High; easily damaged from frequent use	Low
Infrared	An encoded shadow pattern within the card, read using an infrared scanner	Moderate	IR scanners are optical and thus vulnerable to contamination	High
Barium Ferrite	Uses small bits of magnetized barium ferrite placed in a plastic card; the polarity and location of the "spots" determine the coding	Moderate	Low susceptibility to damage; durable since spots are embedded in the material	Moderate–high
Smart cards	Patterns of series of narrow and wide bars and spaces	Short	High susceptibility to damage, low durability	Highest

Source: USEPA (2005).

systems, the vulnerability rating and life-cycle ratings have a reciprocal relationship. For instance, if a given system has a high vulnerability rating, it will almost always have a shorter life cycle.

Card-reader technology can be implemented for facilities of any size and with any number of users. However, because individual systems vary in the complexity of their technology and in the level of security they can provide to a facility, individual users must determine the appropriate system for their needs. Some important features to consider when selecting a card-reader system include:

- The technological sophistication and security level of the card system.
- The size and security needs of the facility.
- The frequency with which the card system will be used. For systems that will experience a high frequency of use, it is important to consider a system that has a longer life cycle and lower vulnerability rating, thus making it more cost effective to implement.
- The conditions in which the system will be used (i.e., will it be used on the interior or exterior of buildings, does it require light or humidity controls, etc.). Most card-reader systems can operate under normal environmental conditions, and therefore this would be a mitigating factor only in extreme conditions
- System costs.

Exterior Intrusion: Buried Sensors

Buried sensors are electronic devices that are designed to detect potential intruders. The sensors are buried along the perimeters of sensitive assets and are able to detect intruder activity both above- and belowground. Some of these systems are composed of individual, stand-alone sensor units, while other sensors consist of buried cables.

There are four types of buried sensors that rely on different types of triggers. These are pressure or seismic, magnetic field, ported coaxial cable, and fiber-optic cables. These four sensors are all covert and terrain following, meaning they are hidden from view and follow the contour of the terrain. The four types of sensors are described in more detail below. Table 8.9 presents the distinctions between the four types of buried sensors.

Table 8.9. Types of Buried Sensors

Type	*Description*
Pressure or seismic	Responds to disturbances in the soil.
Magnetic field	Responds to a change in the local magnetic field caused by the movement of nearby metallic material.
Ported coaxial cables	Responds to motion of a material with a high dielectric constant or high conductivity near the cables.
Fiber-optic cables	Responds to a change in the shape of the fiber that can be sensed using sophisticated sensors and computer signal processing.

Source: Adapted from Garcia (2001).

Exterior Intrusion Sensors

An exterior intrusion sensor is a detection device that is used in an outdoor environment to detect intrusions into a protected area. These devices are designed to detect an intruder and then communicate an alarm signal to an alarm system. The alarm system can respond to the intrusion in many different ways, such as by triggering an audible or visual alarm signal or by sending an electronic signal to a central monitoring location that notifies security personnel of the intrusion.

Intrusion sensors can be used to protect many kinds of assets. Intrusion sensors that protect physical space are classified according to whether they protect indoor or "interior" space (e.g., an entire building or a room within a building) or outdoor or "exterior" space (e.g., a fence line or perimeter). Interior intrusion sensors are designed to protect the interior space of a facility by detecting an intruder who is attempting to enter, or who has already entered, a room or building. In contrast, exterior intrusion sensors are designed to detect an intrusion into a protected outdoor/exterior area. Exterior protected areas are typically arranged as zones or exclusion areas placed so that the intruder is detected early in the intrusion attempt before the intruder can gain access to more valuable assets (e.g., the interior of a building located within the protected area). Early detection creates additional time for security forces to respond to the alarm.

Exterior intrusion sensors are classified according to how the sensor detects the intrusion within the protected area. The three classes of exterior sensor technology include:

- Buried-line sensors
- Fence-associated sensors
- Freestanding sensors

1. Buried-Line Sensors: As the name suggests, buried-line sensors are sensors that are buried underground and are designed to detect disturbances within the ground—such as disturbances caused by an intruder digging, crawling, walking, or running on the monitored ground. Because they sense ground disturbances, these types of sensors are able to detect intruder activity both on the surface and belowground. Individual types of exterior buried-line sensors function in different ways, including by detecting motion, pressure, or vibrations within the protected ground or by detecting changes in some type of field (e.g., magnetic field) that the sensors generate within the protected ground. Specific types of buried-line sensors include pressure or seismic sensors, magnetic field sensors, ported coaxial cables, and fiber-optic cables. Details on each of these sensor types are provided below.
    - *Buried-line pressure* or *seismic sensors* detect physical disturbances to the ground—such as vibrations or soil compression—caused by intruders walking, driving, digging, or otherwise physically contacting the protected ground. These sensors detect disturbances from all directions and therefore can protect an area radially outward from their location; however, because detection may weaken as a function of distance from the disturbance, choosing the correct burial depth for the design area is crucial. In general, sensors buried at a shallow depth protect a relatively small area but have a high probability of detecting intrusion within that area, while sensors buried deeper protect a wider area but have a lower probability of detecting intrusion into that area.
    - *Buried-line magnetic field sensors* detect changes in a local magnetic field that are caused by the movement of metallic objects within that field. This type of sensor can detect ferric metal objects worn or carried by an intruder entering a protected area on foot as well as vehicles being driven into the protected area.
    - *Buried-line ported coaxial cable sensors* detect the motion of any object (e.g., human body, metal, etc.) possessing high conductivity and located within close proximity to the cables. An intruder entering into the protected space creates an active disturbance in the electric field, thereby triggering an alarm condition.
    - *Buried-line fiber-optic cable sensors* detect changes in the attenuation of light signals transmitted within the cable. When the soil around the cable is compressed, the cable is distorted, and the light signal transmitted through the cable changes, initiating an alarm. This type of sensor is easy to install because it can be buried at a shallow depth (only a few centimeters) and still be effective.

2. Fence-Associated Sensors: Fence-associated sensors are either attached to an existing fence or are installed in such a way as to create a fence. These sensors detect disturbances to the fence, such as those caused by an intruder attempting to climb the fence or by an intruder attempting to cut or lift the fence fabric. Exterior fence-associated sensors include fence-disturbance sensors, taut-wire sensor fences, and electric field or capacitance sensors. Details on each of these sensor types are provided below.

    - *Fence-disturbance sensors* detect the motion or vibration of a fence, such as that caused by an intruder attempting to climb or cut through the fence. In general, fence-disturbance sensors are used on chain-link fences or on other fence types where a movable fence fabric is hung between fence posts.
    - *Taut-wire sensor fences* are similar to fence-disturbance sensors except that instead of attaching the sensors to a loose fence fabric, the sensors are attached to a wire that is stretched tightly across the fence. These types of systems are designed to detect changes in the tension of the wire rather than vibrations in the fence fabric. Taut-wire sensor fences can be installed over existing fences or as stand-alone fence systems.
    - *Electric field or capacitance sensors* detect changes in capacitive coupling between wires that are attached to, but electrically isolated from, the fence. As opposed to other fence-associated intrusion sensors, both electric field and capacitance sensors generate an electric field that radiates out from the fence line, resulting in an expanded zone of protection relative to other fence-associated sensors and allowing the sensor to detect the presence of intruders before they arrive at the fence line. *Note:* proper spacing is necessary during installation of the electric field sensor to prevent a would-be intruder from slipping between widely spaced wires.

3. Freestanding Sensors: These sensors, which include active infrared, passive infrared, bistatic microwave, monostatic microwave, dual technology, and video motion detection (VMD) sensors, consist of individual sensor units or components that can be set up in a variety of configurations to meet a user's needs. They are installed aboveground, and depending on how they are oriented relative to each other, they can be used to establish a protected perimeter or a protected space. More details on each of these sensor types are provided below.

    - *Active infrared sensors* transmit infrared energy into the protected space and monitor for changes in this energy caused by intruders entering into that space. In a typical application, an infrared light beam is transmitted from a transmitter unit to a receiver unit. If an intruder crosses the beam, the beam is blocked, and the receiver unit detects a change in the amount of light received, triggering an alarm. Different sensors can see single- or multiple-beam arrays. Single-beam infrared sensors transmit a single infrared beam. In contrast, multiple-beam infrared sensors transmit two or more beams parallel to each other. This multiple-beam sensor arrangement creates an infrared "fence."

- *Passive infrared (PIR) sensors* monitor the ambient infrared energy in a protected area and evaluate changes in that ambient energy that may be caused by intruders moving through the protected area. Detection ranges can exceed one hundred yards on cold days, with size and distance limitations dependent upon the background temperature. PIR sensors generate a nonuniform detection pattern (or "curtain") that has areas (or "zones") of more sensitivity and areas of less sensitivity. The specific shape of the protected area is determined by the detector's lenses. The general shape common to many detection patterns is a series of long "fingers" emanating from the PIR and spreading in various directions. When intruders enter the detection area, the PIR sensor detects differences in temperature due to the intruder's body heat and triggers an alarm. While the PIR leaves unprotected areas between its fingers, an intruder would be detected if he passed from a nonprotected area to a protected area.
- *Microwave sensors* detect changes in received energy generated by the motion of an intruder entering into a protected area. Monostatic microwave sensors incorporate a transmitter and a receiver in one unit, while bistatic sensors separate the transmitter and the receiver into different units. Monostatic sensors are limited to a coverage area of 400 feet, while bistatic sensors can cover an area up to 1,500 feet. For bistatic sensors, a zone of no detection exists in the first few feet in front of the antennas. This distance from the antennas to the point at which the intruder is first detected is known as the offset distance. Due to this offset distance, antennas must be configured so that they overlap one another (as opposed to being adjacent to each other), thereby creating long perimeters with a continuous line of detection.
- *Dual-technology sensors* consist of two different sensor technologies incorporated together into one sensor unit. For example, a dual-technology sensor could consist of a passive infrared detector and a monostatic microwave sensor integrated into the same sensor unit.
- *Video motion detection (VMD) sensors* monitor video images from a protected area for changes in the images. Video cameras are used to detect unauthorized intrusion into the protected area by comparing the most recent image against a previously established one. Cameras can be installed on towers or other tall structures so that they can monitor a large area.

Fences

A fence is a physical barrier that can be set up around the perimeter of an asset. Fences often consist of individual pieces (such as individual pickets in a wooden fence or individual sections of a wrought-iron fence) that are fastened together. Individual sections of the fence are fastened together using posts, which are sunk into the ground to provide stability and strength for the sections of the fence hung between them. Gates are installed between individual sections of the fence to allow access inside the fenced area.

Many fences are used as decorative architectural features to separate physical spaces from each other. They may also be used to physically mark the location of a boundary (such as a fence installed along a property line). However, a fence can also serve as an effective means of physically delaying intruders from gaining access to a critical ESS asset. For example, many utilities install fences around their primary facilities, around remote pump stations, or around hazardous petrochemical materials storage areas or sensitive areas within a facility. Access to the area can be controlled through security at gates or doors through the fence (for example, by posting a guard at the gate or by locking it). In order to gain access to the asset, unauthorized persons would either have to go around or through the fence.

Fences are often compared with walls when determining the appropriate system for perimeter security. While both fences and walls can provide adequate perimeter security, fences are often easier and less expensive to install than walls. However, they do not usually provide the same physical strength that walls do. In addition, many types of fences have gaps between the individual pieces that make up the fence (e.g., the spaces between chain links in a chain-link fence or the spaces between pickets in a picket fence). Thus, many types of fences allow the interior of the fenced area to be seen. This may allow intruders to gather important information about the locations or defenses of vulnerable areas within the facility.

There are numerous types of materials used to construct fences, including chain-link iron, aluminum, wood, or wire. Some types of fences, such as split rails or pickets, may not be appropriate for security purposes because they are traditionally low fences, and they are not physically strong. Potential intruders may be able to easily defeat these fences either by jumping or climbing over them or by breaking through them. For example, the rails in a split-rail fence may be able to be broken easily.

Important security attributes of a fence include the height to which it can be constructed, the strength of the material comprising the fence, the method and strength of attaching the individual sections of the fence together at the posts, and the fence's ability to restrict the view of the assets inside the fence. Additional considerations should include the ease of installing the fence and the

Table 8.10. Comparison of Fence Types

Fence Type	*Height Limitations*	*Strength*	*Installation Requirements*	*Ability to Remove/Reuse*	*Ability to Replace/Repair*
Chain link	12'	Medium	Low	Low	Medium
Iron	12'	High	High	High	High
Wire (wirewall)	12'	High	High	Low	Low
Wood	8'	Low	Low	High	High

Source: USEPA (2005).

ease of removing and reusing sections of the fence. Table 8.10 provides a comparison of the important security and usability features of various fence types.

Some fences can include additional measures to delay, or even detect, potential intruders. Such measures may include the addition of barbed wire, razor wire, or other deterrents at the top of the fence. Barbed wire is sometimes employed at the base of fences as well. This can impede a would-be intruder's progress in even reaching the fence. Fences may also be fitted with security cameras to provide visual surveillance of the perimeter. Finally, some facilities have installed motion sensors along their fences to detect movement on the fence. Several manufacturers have combined these multiple perimeter security features into one product and offer alarms and other security features.

The correct implementation of a fence can make it a much more effective security measure. Security experts recommend the following when a facility constructs a fence:

- The fence should be at least seven to nine feet high.
- Any outriggers, such as barbed wire, that are affixed on top of the fence should be angled out and away from the facility, not in toward the facility. This will make climbing the fence more difficult and will prevent ladders from being placed against the fence.
- Other types of hardware can increase the security of the fence. This can include installing concertina wire along the fence (this can be done in front of the fence or at the top of the fence) or adding intrusion sensors, cameras, or other hardware to the fence.
- All undergrowth should be cleared for several feet (typically six feet) on both sides of the fence. This will allow for a clearer view of the fence by any patrols in the area.
- Any trees with limbs or branches hanging over the fence should be trimmed so that intruders cannot use them to go over the fence. Also, it should be noted that fallen trees can damage fences, so management of trees around the fence may be important. This can be especially important in areas where a fence goes through a remote area.

- Fences that do not block the view from outside the fence to inside the fence allow patrols to see inside the fence without having to enter the facility.
- "No Trespassing" signs posted along fences can be a valuable tool in prosecuting any intruders who claim that the fence was broken and that they did not enter through the fence illegally. Adding signs that highlight the local ordinances against trespassing can further dissuade simple troublemakers from illegally jumping/climbing the fence. Electrical substations and other electrical component installations should have clearly visible signage warning of high voltage and the dangers of electrical shock.

Films for Glass Shatter Protection

Many ESS entities have numerous windows on the outside of buildings, in doors, and in interior offices. In addition, many facilities have glass doors or other glass structures, such as glass walls or display cases. These glass objects are potentially vulnerable to shattering when heavy objects are thrown or launched at them, when explosions occur near them, or when there are high winds (for exterior glass). If the glass is shattered, intruders may potentially enter an area. In addition, shattered glass projected into a room from an explosion or from an object being thrown through a door or window can injure and potentially incapacitate personnel in the room. Materials that prevent glass from shattering can help to maintain the integrity of the door, window, or other glass object and can delay an intruder from gaining access. These materials can also prevent flying glass and thus reduce potential injuries.

Materials designed to prevent glass from shattering include specialized films and coatings. These materials can be applied to existing glass objects to improve their strength and their ability to resist shattering. The films have been tested against many scenarios that could result in glass breakage, including penetration by blunt objects, bullets, high winds, and simulated explosions. Thus, the films are tested against both simulated weather scenarios (which could include both the high winds themselves and the force of objects blown into the glass) and more criminal/terrorist scenarios where the glass is subject to explosives or bullets. Many vendors provide information on the results of these types of tests, and thus potential users can compare different product lines to determine which products best suit their needs.

The primary attributes of films for shatter protection are:

- The materials from which the film is made
- The adhesive that bonds the film to the glass surface
- The thickness of the film

Standard glass safety films are designed from high-strength polyester. Polyester provides both strength and elasticity, which is important in absorbing the impact of an object, spreading the force of the impact over the entire film, and resisting tearing. The polyester is also designed to be resistant to scratching, which can result when films are cleaned with abrasives or other industrial cleaners.

The bonding adhesive is important in ensuring that the film does not tear away from the glass surface. This can be especially important when the glass is broken so that the film does not peel off the glass and allow it to shatter. In addition, films applied to exterior windows can be subject to high concentrations of UV light, which can break down bonding materials.

Film thickness is measured in gauge or mils. According to test results reported by several manufacturers, film thickness appears to affect resistance to penetration or tearing, with thicker films being more resistant. However, the application of a thicker film does not appear to decrease glass fragmentation.

Many manufacturers offer films in different thicknesses. The "standard" film is usually one four-mil layer; thicker films are typically composed of several layers of the standard four-mil sheet. However, newer technologies have allowed the polyester to be "microlayered" to produce a stronger film without significantly increasing its thickness. In this microlayering process, each laminate film is composed of multiple micro-thin layers of polyester woven together at alternating angles. This provides increased strength for the film while maintaining the flexibility and thin profile of one film layer.

As described above, many vendors test their products in various scenarios that would lead to glass shattering, including simulated bomb blasts and simulation of the glass being struck by wind-blown debris. Some manufacturers refer to the General Services Administration standard for bomb blasts, which requires resistance to tearing for a four-pound-per-square-inch (psi) blast. Other manufacturers use other measures and test for resistance to tearing. Many of these tests are not "standard," in that no standard testing or reporting methods have been adopted by any of the accepted standards-setting institutions. However, many of the vendors publish the procedure and the results of these tests on their websites, and this may allow users to evaluate the protectiveness of these films. For example, several vendors evaluate the "protectiveness" of their films and the "hazard" resulting from blasts near windows with and without protective films. Protectiveness is usually evaluated based on the percentage of glass ejected from the window and the height at which that ejected glass travels during the blast (for example, if the blasted glass tends to project upward into a room—potentially toward people's faces—it is a higher hazard than if it is blown downward into the room toward people's feet). There are some standard measures of glass breakage. For example, several vendors indicate that their products exceed the American Society for Testing and Materials (ASTM) standard 64Z-95, "Standard Test Method for Glazing and Glazing Systems Subject to Air Blast Loadings." Vendors often compare the results of some sort of penetration or force test, ballistic test, or simulated explosions with unprotected glass versus glass onto which their films have been applied. Results generally show that applying the films to glass surfaces reduces breakage/penetration of the glass and can reduce the amount of glass ejected from the frame as well as control its direction. This in turn reduces the hazard from flying glass.

In addition to these types of tests, many vendors conduct standard physical tests on their products, such as tests for tensile strength and peel strength. Tensile strength indicates the strength per area of material, while the peel strength indicates the force it would take to peel the product from the glass surface. Several vendors indicate that their products exceed American National Standards Institute (ANSI) standard Z97.1 for tensile strength and adhesion.

Vendors typically have a warranty against peeling or other forms of deterioration of their products. However, the warranty requires that the films be installed by manufacturer-certified technicians to ensure that they are applied correctly and therefore that the warranty is in effect. Warranties from different manufacturers may vary. Some may cover the cost of replacing the material only, while others include material plus installation. Because installation costs are significantly greater than material costs, different warranties may represent large differences in potential costs.

Fire Hydrant Locks

Fire hydrants are installed at strategic locations throughout a community's water distribution system to supply water for firefighting. However, because there are many hydrants in a system and they are often located in residential neighborhoods, industrial districts, and other areas where they cannot be easily observed or guarded, they are potentially vulnerable to unauthorized access. Many municipalities, states, and EPA regions have recognized this potential vulnerability and have instituted programs to lock hydrants. For example, EPA Region 1 has included locking hydrants as number 7 on its "Drinking Water Security and Emergency Preparedness" top-ten list for small groundwater suppliers.

A "hydrant lock" is a physical security device designed to prevent unauthorized access to the water supply through a hydrant. They can also ensure water and water pressure availability to firefighters and prevent water theft and associated lost water revenue. These locks have been successfully used in numerous municipalities and in various climates and weather conditions.

Fire hydrant locks are basically steel covers or caps that are locked in place over the operating nut of a fire hydrant. The lock prevents unauthorized persons from accessing the operating nut and opening the fire hydrant valve. The lock also makes it more difficult to remove the bolts from the hydrant and access the system that way. Finally, hydrant locks shield the valve from being broken off. Should a vandal attempt to breach the hydrant lock by force and succeed in breaking the hydrant lock, the vandal will only succeed in bending the operating valve. If the hydrant's operating valve is bent, the hydrant will not be operational, but the water asset remains protected and inaccessible to vandals. However, the entire hydrant will need to be replaced.

Hydrant locks are designed so that the hydrants can be operated by special "key wrenches" without removing the lock. These specialized wrenches are generally distributed to the fire department, public works department, and other authorized persons so that they can access the hydrants as needed. An inventory of wrenches and their serial numbers is generally kept by a municipality so that the location of all wrenches is known. These operating key wrenches may only be purchased by registered lock owners.

The most important features of hydrants are their strength and the security of their locking systems. The locks must be strong so that they cannot be broken off. Hydrant locks are constructed from stainless or alloyed steel. Stainless-steel locks are stronger and are ideal for all climates; however, they are more expensive than alloy locks. The locking mechanisms for each fire hydrant locking system ensure that the hydrant can only be operated by authorized personnel who have the specialized key to work the hydrant.

Hatch Security

A hatch is basically a door that is installed on a horizontal plane (such as in a floor, a paved lot, or a ceiling) instead of on a vertical plane (such as in a building wall). Hatches are usually used to provide access to assets that are either located underground (such as hatches to basements or underground vaults and storage areas) or above ceilings (such as emergency roof exits). At emergency services facilities, hatches are typically used to provide access to underground vaults containing pumps, meter chambers, valves, or piping, or to the interior of chemical tanks or covered water reservoirs. Securing a hatch by locking it or upgrading materials to give the hatch added strength can help to delay unauthorized access to any asset behind the hatch.

Like all doors, a hatch consists of a frame anchored to the horizontal structure, a door or doors, hinges connecting the door/doors to the frame, and a latching or locking mechanism that keeps the hatch door/doors closed.

It should be noted that improving hatch security is straightforward and that hatches with upgraded security features can be installed new or can be retrofitted for existing applications. Depending on the application, the primary security-related attributes of a hatch are the strength of the door and frame, its resistance to the elements and corrosion, its ability to be sealed against water or gas, and its locking features.

Hatches must be both strong and lightweight so that they can withstand typical static loads (such as people or vehicles walking or driving over them) while still being easy to open.

In addition, because hatches are typically installed at outdoor locations, they are usually made of corrosion-resistant metal that can withstand the elements. Therefore, hatches are typically constructed from high-gauge steel or lightweight aluminum.

Aluminum is typically the material of choice for hatches because it is lightweight and more corrosion resistant than steel. Aluminum is not as rigid as steel, so aluminum hatch doors may be reinforced with aluminum stiffeners to provide extra strength and rigidity. The doors are usually constructed from single or double layers (or "leaves") of material. Single-leaf designs are standard for smaller hatches, while double-leaf designs are required for larger hatches. In addition, aluminum products do not require painting. This is reflected in the warranties available with different products. Product warranties range from ten years to lifetime.

Steel is heavier per square foot than aluminum, and thus steel hatches will be heavier and more difficult to open than aluminum hatches of the same size. However, heavy steel hatch doors may have spring-loaded, hydraulic, or gas openers or other specialized features that help in opening the hatch and in keeping it open.

Many hatches are installed in outdoor areas, often in roadways or pedestrian areas. Therefore, the hatch installed for any given application must be designed to withstand the expected load at that location. Hatches are typically solid to withstand either pedestrian or vehicle loading. Pedestrian loading hatches are typically designed to withstand either 150 or 300 pounds per square feet (psf) of loading. The vehicle loading standard is the American Association of State Highway and Transportation Officials (AASHTO) H-20 wheel loading standard of sixteen thousand pounds over an eight-by-twenty-inch area. It should be noted that these design parameters are for static loads and not dynamic loads; thus, the loading capabilities may not reflect potential resistance to other types of loads that may be more typical of an intentional threat, such as repeated blows from a sledgehammer or pressure generated by bomb blasts or bullets.

The typical design for a watertight hatch includes a channel frame that directs water away from the hatch. This can be especially important in a hatch on a storage tank because this will prevent liquid contaminants from being dumped on the hatch and leaking through into the interior. Hatches can also be constructed with gasket seals that are air-, odor-, and gas-tight.

Typically, hatches for pedestrian loading applications have hinges located on the exterior of the hatch, while hatches designed for H-20 loads have hinges located on the interior of the hatch. Hinges located on the exterior of the hatch may be able to be removed, thereby allowing intruders to remove the hatch door and access the asset behind the hatch. Therefore, installing H-20 hatches even for applications that do not require H-20 loading levels may increase security because intruders will not be able to tamper with the hinges and circumvent the hatch this way. In addition to the location of the hinges, stock hinges can be replaced with heavy-duty or security hinges that are more resistant to tampering.

The hatch locking mechanism is perhaps the most important part of hatch security. There are a number of locks that can be implemented for hatches, including:

- Slam locks (internal locks that are located within the hatch frame)
- Recessed cylinder locks
- Bolt locks
- Padlocks

Ladder Access Control

Emergency services facilities have a number of assets that are raised above ground level, including electrical substations, transmitting stations, raised conduit systems, and roof access points into buildings. In addition, communications equipment, antennas, or other electronic devices may be located on the top of these raised assets. Typically, these assets are reached by ladders that are permanently anchored to the asset. For example, raised petrochemical/water tanks are typically accessed by ladders that are bolted to one of the legs of the tank. Controlling access to these raised assets by controlling access to the ladder can increase security at an emergency services facility.

A typical ladder access control system consists of some type of cover that is locked or secured over the ladder. The cover can be a casing that surrounds most of the ladder or a door or shield that covers only part of the ladder. In either case, several rungs of the ladder (the number of rungs depends on the size of the cover) are made inaccessible by the cover, and these rungs can only be accessed by opening or removing the cover. The cover is locked so that only authorized personnel can open or remove it and use the ladder. Ladder access controls are usually installed several feet above ground level, and they usually extend several feet up the ladder so that they cannot be circumvented by someone accessing the ladder above the control system.

The important features of ladder access control are the size and strength of the cover and its ability to lock or otherwise be secured from unauthorized access. The covers are constructed from aluminum or some type of steel. This should provide adequate protection from being pierced or cut through. The metals are corrosion resistant so that they will not corrode or become fragile from extreme weather conditions in outdoor applications. The bolts used to install each of these systems are galvanized steel. In addition, the bolts for each cover are installed on the inside of the unit so they cannot be removed from the outside.

Locks

A lock is a type of physical security device that can be used to delay or prevent a door, gate, window, manhole, filing cabinet drawer, or some other physical feature from being opened, moved, or operated. Locks typically work by connecting two pieces together—such as by connecting a door to a doorjamb or a manhole to its casement. Every lock has two modes—engaged (or "locked") and disengaged (or "opened"). When a lock is disengaged, the asset on which the lock is installed can be accessed by anyone, but when the lock is engaged, only by those authorized to access to the locked asset.

Before discussing locks and their applicability, it is important to discuss key control. Based on our experience, many emergency services facilities (and others) have no idea how many keys for various site/equipment locks have been issued to employees over the years. Many facilities simply issue keys to employees at hiring, with no accountability for the keys upon the employee's departure. Needless to say, this is not good security policy. You can have the best-made locks available installed throughout your facilities, but if you do not have proper key control, you do not have proper security.

Locks are excellent security features because they have been designed to function in many ways and to work on many different types of assets. Locks can also provide different levels of security depending on how they are designed and implemented. The security provided by a lock is dependent on several factors, including its ability to withstand physical damage (i.e., can it be cut off, broken, or otherwise physically disabled) as well as its requirements for supervision or operation (i.e., combinations may need to be changed frequently so that they are not compromised and the locks remain secure). While there is no single definition of the "security" of a lock, locks are often described as minimum, medium, or maximum security. Minimum-security locks are those that can be easily disengaged (or "picked") without the correct key or code, or those that can be disabled easily (such as small padlocks that can be cut with bolt cutters). Higher-security locks are more complex and thus are more difficult to pick, or are sturdier and more resistant to physical damage.

Many locks, such as many door locks, only need to be unlocked from one side. For example, most door locks need a key to be unlocked only from the outside. A person opens such devices, called single-cylinder locks, from the inside by pushing a button or by turning a knob or handle. Double-cylinder locks require a key to be locked or unlocked from both sides.

Manhole Intrusion Sensors

Manholes are commonly found in emergency services sites. Manholes are designed to provide access to underground utilities, meter vaults, petrochemical pumping rooms, etc., and therefore they are potential entry points to a system. Because many utilities run under other infrastructure (roads, buildings), manholes also provide potential access points to critical infrastructure as well as petrochemical process assets. In addition, because the portion of the system to which manholes provide entry is primarily located underground, access to a system through a manhole increases the chance that an intruder will not be seen. Therefore, protecting manholes can be a critical component of guarding an entire plant site and a surrounding community.

There are multiple methods for protecting manholes, including preventing unauthorized personnel from physically accessing the manhole and detecting attempts at unauthorized access to the manhole.

A manhole intrusion sensor is a physical security device designed to detect unauthorized access to the facility through a manhole. Monitoring a manhole that provides access to a chemical plant or processing system can mitigate two distinct types of threats. First, monitoring a manhole may detect access of unauthorized personnel to chemical systems or assets through the manhole. Second, monitoring manholes may also allow the detection of intruders attempting to place explosive or other destructive (WMD) devices into the petrochemical system.

Several different technologies have been used to develop manhole intrusion sensors, including mechanical systems, magnetic systems, and fiber-optic or infrared sensors. Some of these intrusion sensors have been specifically designed for manholes, while others consist of standard, off-the-shelf intrusion sensors that have been implemented in a system specifically designed for application in a manhole.

Manhole Locks

A "manhole lock" is a physical security device designed to delay unauthorized access to the emergency services facility or system through a manhole.

Passive Security Barriers

One of the most basic threats facing any facility is from intruders accessing the facility with the intention of causing damage to its assets. These threats may include intruders actually entering the facility, as well as intruders attacking the facility from outside without actually entering it (e.g., detonating a bomb near enough to the facility to cause damage within its boundaries).

Security barriers are one of the most effective ways to counter the threat of intruders accessing a facility or the facility perimeter. Security barriers are large, heavy structures that are used to control access through a perimeter by either vehicles or personnel. They can be used in many different ways depending on how/where they are located at the facility. For example, security barriers can be used on or along driveways or roads to direct traffic to a checkpoint (e.g., a facility may install Jersey barriers in a road to direct traffic in a certain direction). Other types of security barriers (crash beams, gates) may be installed at the checkpoint so that guards can regulate which vehicles can access the facility. Finally, other security barriers (e.g., bollards or security planters) can be used along the facility perimeter to establish a protective buffer area between the facility and approaching vehicles. Establishing such a protective buffer can help in mitigating the effects of the type of bomb blast described above, both by potentially absorbing some of the blast and by increasing the "standoff" distance between the blast and the facility (the force of an explosion is reduced as the shock wave travels farther from the source, and thus the farther the explosion is from the target, the less effective it will be in damaging the target).

Security barriers can be either "active" or "passive." Active barriers, which include gates, retractable bollards, wedge barriers, and crash barriers, are readily movable, and thus they are typically used in areas where they must be moved often to allow vehicles to pass—such as in roadways at entrances and exits to a facility. In contrast to active security barriers, passive security barriers, which include Jersey barriers, bollards, and security planters, are not designed to be moved on a regular basis, and thus they are typically used in areas where access is not required or allowed, such as along building perimeters or in traffic control areas. Passive security barriers are typically large, heavy structures that are usually several feet high, and they are designed so that even heavy-duty vehicles cannot go over or through them. Therefore, they can be placed in a roadway parallel to the flow of traffic so that they direct traffic in a certain direction (such as to a guardhouse, a gate, or some other sort of checkpoint), or perpendicular to traffic such that they prevent a vehicle from using a road or approaching a building or area.

Security for Doorways: Side-Hinged Doors

Doorways are the main access points to a facility or to rooms within a building. They are used on the exterior or in the interior of buildings to provide privacy and security for the areas behind them. Different types of doorway security systems may be installed in different doorways depending on the requirements of the buildings or rooms. For example, exterior doorways tend to have heavier doors to withstand the elements and to provide some security to the entrance of the building. Interior doorways in office areas may have lighter doors that are primarily designed to provide privacy rather than security. Therefore, these doors may be made of glass or lightweight wood. Doorways in industrial areas may have sturdier doors than do other interior doorways and may be designed to provide protection or security for areas behind the doorway. For example, fireproof doors may be installed in chemical storage areas or in other areas where there is a danger of fire.

Because they are the main entries into a facility or a room, doorways are often prime targets for unauthorized entry into a facility or an asset. Therefore, securing doorways may be a major step in providing security at a facility.

A doorway includes four main components:

- The door, which blocks the entrance. The primary threat to the actual door is breaking or piercing through the door. Therefore, the primary security features of doors are their strength and resistance to various physical threats, such as fire or explosions.
- The door frame, which connects the door to the wall. The primary threat to a door frame is that the door can be pried away from the frame. Therefore, the primary security feature of a door frame is its resistance to prying.
- The hinges, which connect the door to the door frame. The primary threat to door hinges is that they can be removed or broken, which will allow intruders to remove the entire door. Therefore, security hinges are designed to be resistant to breaking. They may also be designed to minimize the threat of removal from the door.
- The lock, which connects the door to the door frame. Use of the lock is controlled through various security features, such as keys, combinations, etc., such that only authorized personnel can open the lock and go through the door. Locks may also incorporate other security features,

such as software or other systems to track overall use of the door or to track individuals using the door, etc.

Each of these components is integral in providing security for a doorway, and upgrading the security of only one of these components while leaving the other components unprotected may not increase the overall security of the doorway. For example, many facilities upgrade door locks as a basic step in increasing the security of a facility. However, if the facilities do not also focus on increasing security for the door hinges or the door frame, the door may remain vulnerable to being removed from its frame, thereby defeating the increased security of the door lock.

The primary attribute for the security of a door is its strength. Many security doors are four- to twenty-gauge hollow metal doors consisting of steel plates over a hollow cavity reinforced with steel stiffeners to give the door extra stiffness and rigidity. This increases resistance to blunt force used to try to penetrate through the door. The space between the stiffeners may be filled with specialized materials to provide fire, blast, or bullet resistance to the door.

The Window and Door Manufacturers Association has developed a series of performance attributes for doors. These include:

- Structural resistance
- Forced entry resistance
- Hinge-style screw resistance
- Split resistance
- Hinge resistance
- Security rating
- Fire resistance
- Bullet resistance
- Blast resistance

The first five bullet points provide information on a door's resistance to standard physical breaking and prying attacks. These tests are used to evaluate the strength of the door and the resistance of the hinges and the frame in a standardized way. For example, the rack load test simulates a prying attack on a corner of the door. A test panel is restrained at one end, and a third corner is supported. Loads are applied and measured at the fourth corner. The door impact test simulates a battering attack on a door and frame using impacts of two hundred foot-pounds by a steel pendulum. The door must remain fully operable after the test. It should be noted that door glazing is also rated for resistance to shattering, etc. Manufacturers will be able to provide security ratings for these features of a door as well.

Door frames are an integral part of doorway security because they anchor the door to the wall. Door frames are typically constructed from wood or steel, and they are installed such that they extend for several inches over the doorway that has been cut into the wall. For added security, frames can be designed to have varying degrees of overlap with, or wrapping over, the underlying wall. This can make prying the frame from the wall more difficult. A frame formed from a continuous piece of metal (as opposed to a frame constructed from individual metal pieces) will prevent prying between pieces of the frame.

Many security doors can be retrofitted into existing frames; however, many security door installations include replacing the door frame as well as the door itself. For example, bullet resistance per Underwriters Laboratories (UL) 752 requires resistance of the door and frame assembly, and thus replacing the door only would not meet UL 752 requirements.

Valve Lockout Devices

Valves are utilized as control elements in compressed gas, fuel oil/natural gas, and petrochemical process piping networks and for some firefighting chemicals. They regulate the flow of both liquids and gases by opening, closing, or obstructing a flow passageway. Valves are typically located where flow control is necessary. They can be located in-line or at pipeline or tank entrance and exit points. They can serve multiple purposes in a process pipe network, including:

- Redirecting and throttling flow
- Preventing backflow
- Shutting off flow to a pipeline or tank (for isolation purposes)
- Releasing pressure
- Draining extraneous liquid from pipelines or tanks
- Introducing chemicals into the process network
- As access points for sampling process water

Valves may be located either aboveground or belowground. It is critical to provide protection against valve tampering. For example, tampering with a pressure relief valve could result in a pressure buildup and potential explosion in the piping network. On a larger scale, the addition of a contaminant or non-compatible chemical substance to the chemical processing system through an unprotected valve could result in the catastrophic release of that contaminant to the general population.

Different security products are available to protect aboveground versus belowground valves. For example, valve lockout devices can be purchased to protect valves and valve controls located aboveground. Vaults containing underground valves can be locked to prevent access to those valves.

As described above, a lockout device can be used as a security measure to prevent unauthorized access to aboveground valves located within petrochemical processing systems. Valve lockout devices are locks that are specially designed to fit over valves and valve handles to control their ability to be turned or seated. These devices can be used to lock the valve into the desired position. Once the valve is locked, it cannot be turned unless the locking device is unlocked or is removed by an authorized individual.

Various valve lockout options are available for industrial use, including:

- Cable lockouts
- Padlocked chains/cables
- Valve-specific lockouts

Many of these lockout devices are not specifically designed for use in the emergency services sector (e.g., chains, padlocks) but are available from a local hardware store or manufacturer specializing in safety equipment. Other lockout devices (for example, valve-specific lockouts or valve box locks) are more specialized and must be purchased from safety or valve-related equipment vendors.

The three most common types of valves for which lockout devices are available are gate, ball, and butterfly valves. Each is described in more detail below.

- Gate Valve Lockouts: These are designed to fit over the operating hand wheel of the gate valve to prevent it from being turned. The lockout is secured in place with a padlock. Two types of gate valve lockouts are available: diameter-specific and adjustable. Diameter-specific lockouts are available for handles ranging from one inch to thirteen inches in diameter. Adjustable gate valve lockouts can be adjusted to fit any handle ranging from one inch to six or more inches in diameter.
- Ball Valve Lockouts: There are several different configurations available to lock out ball valves, all of which are designed to prevent rotation of the valve handle. The three major configurations available are a wedge shape for one-inch to three-inch valves, a lockout that completely covers ⅜ inch

to eight-inch ball valve handles, and a universal lockout that can be applied to quarter-turn valves of varying sizes and geometric handle dimensions. All three types of ball valve lockouts can be installed by sliding the lockout device over the ball valve handle and securing it with a padlock.

- Butterfly Valve Lockouts: The butterfly valve lockout functions in a similar manner to the ball valve lockout. The polypropylene lockout device is placed over the valve handle and secured with a padlock. This type of lockout has been commonly used in the bottling industry.

A major difference between valve-specific lockout devices and the padlocked chain or cable lockouts discussed earlier is that they do not need to be secured to an anchoring device in the floor or the piping system. In addition, valve-specific lockouts eliminate potential tripping or access hazards that may be caused by chains or cable lockouts applied to valves located near walkways or frequently maintained equipment.

Valve-specific lockout devices are available in a variety of colors, which can be useful in distinguishing different valves. For example, different-colored lockouts can be used to distinguish the type of liquid passing through the valve (e.g., treated, untreated, potable, petrochemical) or to identify the party responsible for maintaining the lockout. Implementing a system of different-colored locks on operating valves can increase system security by reducing the likelihood of an operator inadvertently opening the wrong valve and causing a problem in the system.

Security for Vents

Vents are installed in some aboveground communications sector storage areas to allow safe venting of off-gases. The specific vent design for any given application will vary depending on the design of the chemical storage vessel. However, every vent consists of an open-air connection between the storage container and the outside environment. Although these air exchange vents are an integral part of covered or underground chemical storage containers, they also represent a potential security threat. Improving vent security by making the vents tamper resistant or by adding other security features, such as security screens or security covers, can enhance the security of the entire petrochemical processing system.

Many municipalities already have specifications for vent security at their local chemical industrial assets. These specifications typically include the following requirements:

- Vent openings are to be angled down or shielded to minimize the entrance of surface and/or rainwater into the vent through the opening.
- Vent designs are to include features to exclude insects, birds, animals, and dust.
- Corrosion-resistant materials are to be used to construct the vents.

Visual Surveillance Monitoring

Visual surveillance is used to detect threats through continuous observation of important or vulnerable areas of an asset. The observations can also be recorded for later review or use (for example, in court proceedings). Visual surveillance systems can be used to monitor various parts of production, distribution, or pumping/compressing systems, including the perimeter of a facility, outlying pumping stations, or entry or access points into specific buildings. These systems are also useful in recording individuals who enter or leave a facility, thereby helping to identify unauthorized access. Images can be transmitted live to a monitoring station, where they can be monitored in real time, or they can be recorded and reviewed later. Many emergency services facilities have found that a combination of electronic surveillance and security guards provides an effective means of facility security.

Visual surveillance is provided through a closed-circuit television (CCTV) system in which the capture, transmission, and reception of an image is localized within a closed "circuit." This is different from other broadcast images, such as over-the-air television, which is broadcast over the air to any receiver within range.

At a minimum, a CCTV system consists of:

- One or more cameras
- A monitor for viewing the images
- A system for transmitting the images from the camera to the monitor

Specific attributes and features of camera systems, lenses, and lighting systems are presented in table 8.11.

Communication Integration

In this section, those devices necessary for communication and integration of ESS processing and response operations, such as electronic controllers, two-way radios, and wireless data communications, are discussed. In regard to security applications, electronic controllers are used to automatically activate equipment (such as lights, surveillance cameras, audible alarms, or locks)

Table 8.11. **Attributes of Camera Systems, Lenses, and Lighting Systems**

Attribute	*Discussion*
Camera Systems	
Camera Type	Major factors in choosing the correct camera are the resolution of the image required and lighting of the area to be viewed. **Solid state** (including charge-coupled devices, charge-priming devices, charge-injection devices, and metal oxide substrates)—These cameras are becoming predominant in the marketplace because of their high resolution and their elimination of problems inherent in tube cameras. **Thermal**—These cameras are designed for night vision. They require no light and use differences in temperature between objects in the field of view to produce a video image. Resolution is low compared to other cameras, and the technology is currently expensive relative to other technologies. **Tube**—These cameras can provide high resolution but are susceptible to burnout and must be replaced after 1–2 years. In addition, tube performance can degrade over time. Finally, tube cameras are prone to burn images in the tube replacement.
Resolution (the ability to see fine details)	User must determine the amount of resolution required depending on the level of detail required for threat determination. A high-definition focus with a wide field of vision will give an optimal viewing area.
Field-of-vision width	Cameras are designed to cover a defined field of vision, which is usually defined in degrees. The wider the field of vision, the more area a camera will be able to monitor.
Type of image produced (color, black and white, thermal)	Color images may allow the identification of distinctive markings, while black-and-white images may provide sharper contrast. Thermal imaging allows the identification of heat sources (such as human beings or other living creatures) from low-light environments; however, thermal images are not effective in identifying specific individuals (i.e., for subsequent legal processes).
Pan/tilt/zoom (PTZ)	Panning (moving the camera in a horizontal plane), tilting (moving the camera in a vertical plane), and zooming (moving the lens to focus on objects that are at different distances from the camera) allow the camera to follow a moving object. Different systems allow these functions to be controlled manually or automatically. Factors to be considered in PTZ cameras are the degree of coverage for pan and tilt functions and the power of the zoom lens.

(*continued*)

Table 8.11. *Continued*

Attribute	*Discussion*
Lenses	
Format	Lens format determines the maximum image size to be transmitted.
Focal length	This is the distance from the lens to the center of the focus. The greater the focal length, the higher the magnification, but the narrower the field of vision.
F-number	F-number is the ability to gather light. Smaller f-numbers may be required for outdoor applications where light cannot be controlled as easily.
Distance and width approximation	The distance and width approximations are used to determine the geometry of the space that can be monitored at the best resolution.
Lighting Systems	
Intensity	Light intensity must be great enough for the camera type to produce sharp images. Light can be generated from natural or artificial sources. Artificial sources can be controlled to produce the amount and distribution of light required for a given camera and lens.
Evenness	Light must be distributed evenly over the field of view so that there are no darker or shadowy areas. If there are lighter vs. darker areas, brighter areas may appear washed out (i.e., details cannot be distinguished), while no specific objects can be viewed in the darker areas.
Location	Light sources must be located above the camera so that light does not shine directly into the camera.

Source: USEPA (2005).

when they are triggered. Triggering could be in response to a variety of scenarios, including the tripping of an alarm or motion sensor, the breaking of a window or a glass door, variation in vibration sensor readings, or simply through input from a timer.

Two-way wireless radios allow two or more users that have their radios tuned to the same frequency to communicate instantaneously with each other without the radios being physically linked together with wires or cables.

Wireless data communications devices are used to enable transmission of data between computer systems and/or between a SCADA server and its sensing devices, without individual components being physically linked together via wires or cables. In industrial petrochemical processing systems, these devices are often used to link remote monitoring stations (e.g., SCADA components) or portable computers (e.g., laptops) to computer networks without using physical wiring connections.

Electronic Controllers

An electronic controller is a piece of electronic equipment that receives incoming electric signals and uses preprogrammed logic to generate electronic output signals based on the incoming signals. While electronic controllers can be implemented for any application that involves inputs and outputs (for example, control of a piece of machinery in a factory), in a security application, these controllers essentially act as the system's "brain" and can respond to specific security-related inputs with preprogrammed output responses. These systems combine the control of electronic circuitry with a logic function such that circuits are opened and closed (and thus equipment is turned on and off) through some preprogrammed logic. The basic principle behind the operation of an electrical controller is that it receives electronic inputs from sensors or any device generating an electrical signal (for example, electrical signals from motion sensors) and then uses its preprogrammed logic to produce electrical outputs (for example, these outputs could turn on power to a surveillance camera or to an audible alarm). Thus, these systems automatically generate a preprogrammed, logical response to a preprogrammed input scenario.

The three major types of electronic controllers are timers, electromechanical relays, and programmable logic controllers (PLCs), which are often called "digital relays." Each of these types of controller is discussed in more detail below.

Timers use internal signals/inputs (in contrast to externally generated inputs) and generate electronic output signals at certain times. More specifically, timers control electric current flow to any application to which they are connected, and they can turn the current on or off on a schedule prespecified by the user. Typical timer range (amount of time that can be programmed to elapse before the timer activates linked equipment) is from 0.2 seconds to ten hours, although some of the more advanced timers have ranges of up to sixty hours. Timers are useful in fixed applications that don't require frequent schedule changes. For example, a timer can be used to turn on the lights in a room or building at a certain time every day. Timers are usually connected to their own power supply (usually 120–240 volts).

In contrast to timers, which have internal triggers based on a regular schedule, electromechanical relays and PLCs have both external inputs and external outputs. However, PLCs are more flexible and more powerful than electromechanical relays, and thus this section focuses primarily on PLCs as the predominant technology for security-related electronic control applications.

Electromechanical relays are simple devices that use a magnetic field to control a switch. Voltage applied to the relay's input coil creates a magnetic field, which attracts an internal metal switch. This causes the relay's contacts to touch, closing the switch and completing the electrical circuit. This activates any linked equipment. These types of systems are often used for high-voltage applications, such as in some automotive and other manufacturing processes.

Two-Way Radios

Two-way radios, as discussed here, are limited to a direct unit-to-unit radio communication, either via single unit-to-unit transmission and reception or via multiple handheld units to a base station radio contact and distribution system. Radio frequency spectrum limitations apply to all handheld units and are directed by the FCC. This also distinguishes a handheld unit from a base station or base station unit (such as those used by an amateur [ham] radio operator), which operate under different wavelength parameters.

Two-way radios allow a user to contact another user or group of users instantly on the same frequency and to transmit voice or data without the need for wires. They use "half-duplex" communications, or communications that can only be transmitted or received; they cannot transmit and receive simultaneously. In other words, only one person may talk, while other personnel with radio(s) can only listen. To talk, the user depresses the talk button and speaks into the radio. The radio then transmits the voice wirelessly to the receiving radios. When the speaker has finished speaking and the channel has cleared, users on any of the receiving radios can transmit, either to answer the first transmission or to begin a new conversation. In addition to carrying voice data, many types of wireless radios also allow the transmission of digital data, and these radios may be interfaced with computer networks that can use or track these data. For example, some two-way radios can send information such as GPS data or the ID of the radio. Some two-way radios can also send data through a SCADA system.

Wireless radios broadcast these voice or data communications over the airwaves from the transmitter to the receiver. While this can be an advantage in that the signal emanates in all directions and does not need a direct physical connection to be received at the receiver, it can also make the communications vulnerable to being blocked, intercepted, or otherwise altered. However, security features are available to ensure that the communications are not tampered with.

Wireless Data Communications

A wireless data communication system consists of two components, a wireless access point (WAP) and a wireless network interface card (sometimes also referred to as a "client"), which work together to complete the communications link. These wireless systems can link electronic devices, computers, and computer systems together using radio waves, thus eliminating the need for these individual components to be directly connected together through physical wires.

The WAP provides wireless data communications service. It usually consists of a housing (which is constructed from plastic or metal depending on the environment it will be used in) containing a circuit board, flash memory that holds software, one of two external ports to connect to existing wired networks, a wireless radio transmitter/receiver, and one or more antenna connections. Typically, the WAP requires a one-time user configuration to allow the device to interact with the local area network (LAN). This configuration is usually done via a web-driven software application that is accessed via a computer.

A wireless network interface card/client is a piece of hardware that is plugged into a computer and enables that computer to make a wireless network connection. The card consists of a transmitter, functional circuitry, and a receiver for the wireless signal, all of which work together to enable communication between the computer, its wireless transmitter/receiver, and its antenna connection. Wireless cards are installed in a computer through a variety of connections, including USB adapters, laptop CardBus (PCMCIA) card, or desktop peripheral (PCI) cards. As with the WAP, software is loaded onto the user's computer, allowing configuration of the card so that it may operate over the wireless network.

Two of the primary applications for wireless data communications systems are to enable mobile or remote connections to a LAN and to establish wireless communications links between SCADA remote terminal units (RTUs) and sensors in the field. Wireless card connections are usually used for LAN access from mobile computers. Wireless cards can also be incorporated into RTUs to allow them to communicate with sensing devices that are located remotely.

Cyber Protection Devices

Various cyber protection devices are currently available for use in protecting ESS computer systems. These protection devices include antivirus and pest-eradication software, firewalls, and network intrusion hardware/software. These products are discussed in this section.

Antivirus and Pest-Eradication Software

Antivirus programs are designed to detect, delay, and respond to programs or pieces of code that are specifically designed to harm computers. These programs are known as "malware." Malware can include computer viruses, worms, and Trojan horse programs (programs that appear to be benign but which have hidden harmful effects).

Pest-eradication tools are designed to detect, delay, and respond to "spyware" (strategies that websites use to track user behavior, such as by sending "cookies" to the user's computer) and hacker tools that track keystrokes (keystroke loggers) or passwords (password crackers).

Viruses and pests can enter a computer system through the internet or through infected floppy disks or CD-ROMs. They can also be placed onto a system by insiders. Some of these programs, such as viruses and worms, then move within a computer's drives and files, or between computers if the computers are networked to each other. This malware can deliberately damage files, utilize memory and network capacity, crash application programs, or initiate transmissions of sensitive information from a computer. While the specific mechanisms of these programs differ, they can infect files and even the basic operating program of the computer firmware/hardware.

The most important features of an antivirus program are its abilities to identify potential malware and alert a user before infection occurs, as well as its ability to respond to a virus already resident on a system. Most of these programs provide a log so that the user can see what viruses have been detected and where they were detected. After detecting a virus, the antivirus software may delete the virus automatically, or it may prompt the user to delete the virus. Some programs will also fix files or programs damaged by the virus.

Various sources of information are available to inform the general public and computer system operators about new viruses being detected. Since antivirus programs use signatures (or snippets of code or data) to detect the presence of a virus, periodic updates are required to identify new threats. Many antivirus software providers offer free upgrades that are able to detect and respond to the latest viruses.

Firewalls

A firewall is an electronic barrier designed to keep computer hackers, intruders, or insiders from accessing specific data files and information on an ESS's computer network or other electronic/computer systems. Firewalls operate by evaluating and then filtering information coming through a public network (such as the internet) into the utility's computer or other electronic system. This evaluation can include identifying source or destination addresses and ports and allowing or denying access based on this identification.

There are two methods used by firewalls to limit access from the public network to the utility's computers or other electronic systems:

- The firewall may deny all traffic unless it meets certain criteria.
- The firewall may allow all traffic through unless it meets certain criteria.

A simple example of the first method is to screen requests to ensure that they come from an acceptable (i.e., previously identified) domain name and Internet Protocol address. Firewalls may also use more complex rules that analyze the application data to determine if the traffic should be allowed through. For example, the firewall may require user authentication (e.g., use of a password) to access the system. How a firewall determines what traffic to let through depends on which network layer it operates at and how it is configured. Some of the pros and cons of various methods to control traffic flowing in and out of a network are provided in table 8.12.

Firewalls may be a piece of hardware, a software program, or an appliance card that contains both. Advanced features that can be incorporated into firewalls allow for the tracking of attempts to log on to the LAN system. For example, a report of successful and unsuccessful log-in attempts may be generated for the computer specialist to analyze. For systems with mobile users, firewalls allow remote access into the private network by the use of secure log-on procedures and authentication certificates. Most firewalls have a graphical user interface for managing the firewall.

In addition, new Ethernet firewall cards that fit in the slot of an individual computer can bundle additional layers of defense (like encryption and permit/deny) for individual computer transmissions to the network interface function. These new cards have only a slightly higher cost than traditional network interface cards.

Network Intrusion Hardware/Software

Network intrusion detection and prevention systems are software- and hardware-based programs designed to detect unauthorized attacks on a computer network system.

While other applications, such as firewalls and antivirus software, share similar objectives with network intrusion systems, network intrusion systems provide a deeper layer of protection beyond the capabilities of these other systems because they evaluate patterns of computer activity rather than specific files.

Table 8.12. Pros and Con of Various Firewall Methods for Controlling Network Access

Method	*Description*	*Pros*	*Cons*
Packet filtering	Incoming and outgoing packets (small chunks of data) are analyzed against a set of filters. Packets that make it through the filters are sent to the requesting system, and all others are discarded. There are two types of packet filter: static (the most common) and dynamic.	Static filtering is relatively inexpensive, and little maintenance is required; well suited for closed environments where access to or from multiple addresses is not allowed.	Leaves permanent open holes in the network; allows direct connection to internal hosts by external sources; offers no user authentication; method can be unmanageable in large networks.
Proxy service	Information from the internet is retrieved by the firewall and then sent to the requesting system and vice versa; in this way, the firewall can limit the information made known to the requesting system, making vulnerabilities less apparent.	Only allows temporary open holes in the network perimeter; can be used for all types of internal protocol services.	Allows direct connections to internal hosts by external clients; offers no user authentication.
Stateful pattern recognition	This method examines and compares the contents of certain key parts of an information packet against a database of acceptable information. Information traveling from inside the firewall to the outside is monitored for specific defining characteristics, then incoming information is compared to these characteristics; if the comparison yields a reasonable match, the information is allowed through; if not, the information is discarded.	Provides a limited time window to allow packets of information to be sent; does not allow any direct connections between internal and external hosts; supports user-level authentication.	Slower than packet filtering; does not support all types of connections.

Source: USEPA (2005).

It is worth noting that attacks may come from either outside or within the system (i.e., from an insider), and network intrusion detection systems may be more applicable for detecting patterns of suspicious activity from inside a facility (i.e., accessing sensitive data, etc.) than are other information technology solutions.

Network intrusion detection systems employ a variety of mechanisms to evaluate potential threats. The types of search and detection mechanisms are dependent upon the level of sophistication of the system. Some of the available detection methods include:

- Protocol Analysis: Protocol analysis is the process of capturing, decoding, and interpreting electronic traffic. The protocol analysis method of network intrusion detection involves the analysis of data captured during transactions between two or more systems or devices and the evaluation of these data to identify unusual activity and potential problems. Once a problem is isolated and recorded, problems or potential threats can be linked to pieces of hardware or software. Sophisticated protocol analysis will also provide statistics and trend information on the captured traffic.
- Traffic Anomaly Detection: Traffic anomaly detection identifies potential threatening activity by comparing incoming traffic to "normal" traffic patterns and identifying deviations. It does this by comparing user characteristics against thresholds and triggers defined by the network administrator. This method is designed to detect attacks that span a number of connections rather than a single session.
- Network Honeypot: This method establishes nonexistent services in order to identify potential hackers. A network honeypot impersonates services that don't exist by sending fake information to people scanning the network. It identifies the attacker when they attempt to connect to the service. There is no reason for legitimate traffic to access these resources because they don't exist; therefore any attempt to access them constitutes an attack.
- Anti-intrusion Detection System Evasion Techniques: These methods are designed by attackers who may be trying to evade intrusion detection system scanning. They include methods called IP defragmentation, TCP stream reassembly, and deobfuscation.

While these detection systems are automated, they can only indicate patterns of activity, and a computer administrator or other experienced individual must interpret activities to determine whether or not they are potentially harmful. Monitoring the logs generated by these systems can be time consuming, and there may be a learning curve to determine a baseline of "normal" traffic patterns from which to distinguish potential suspicious activity.

NOTE

1. It is important to point out that even though the following USEPA security asset and device recommendations were first made for the water/wastewater critical infrastructure, these recommendations are applicable to all other critical infrastructure sectors, including the defense industrial base (DIB) sector.

REFERENCES AND RECOMMENDED READING

Garcia, M. L. (2001). *The Design and Evaluation of Physical Protection Systems*. Boston: Butterworth-Heinemann.

IBWA. (2004). *Bottled Water Safety and Security*. Alexandria, VA: International Bottled Water Association.

NAERC. (2002). *Security Guidelines for the Electricity Sector*. Washington, DC: North American Electric Reliability Council.

Schneier, B. (2000). *Secrets & Lies*. New York: Wiley.

USEPA. (2005). *Water and Wastewater Security Product Guide*. http://cfpub.epa.gov.safewater/watersecurity/guide, accessed April 14, 2016.

Chapter 9

The Paradigm Shift

The events of 9/11 dramatically changed this nation and focused us on combating terrorism. As a result, in 2003 and subsequent years, the Department of Homeland Security (DHS), in conjunction with members from the general public, state and local agencies, and private groups concerned with the safety of critical infrastructures, established a Water Security Working Group (WSWG) to consider and make recommendations on infrastructure security issues. Although initially created to make recommendations for water/wastewater security, WSWG is an excellent template for use with other critical infrastructures, including emergency services assets. For example, the WSWG identified active and effective security practices for critical infrastructure and provided an approach for adopting these practices. It also recommended mechanisms to provide incentives that facilitate broad and receptive response among critical infrastructure sectors to implement active and effective security practices. Finally, WSWG recommended mechanisms to measure progress and achievements in implementing active and effective security practices, and to identify barriers to implementation.

The WSWG recommendations on security are structured to maximize benefits to critical industries by emphasizing actions that have the potential both to improve the quality or reliability of service and to enhance security. These recommendations, based on original recommendations from the 2003 National Drinking Water Advisor Council (NDWAC), were designed primarily, as the name suggests, for use by water systems of all types and sizes, including systems that serve fewer than 3,300 people. However, it is the author's opinion, based on personal experience, that NDWAC's recommendations, when properly adapted to applicable circumstances and locations, can be applied to any and all critical infrastructure sectors, including the emergency services sector.

The NDWAC identified fourteen features of active and effective security programs that are important for increasing security and are relevant across the broad range of utility circumstances and operating conditions. USEPA (2003) points out that the fourteen features are, in many cases, consistent with the steps needed to maintain technical, management, and operational performance capacity related to overall water quality; as pointed out earlier, these steps can be applied to other critical infrastructures as well. Many facilities may be able to adopt some of the features with minimal, if any, capital investment.

FOURTEEN FEATURES OF ACTIVE AND EFFECTIVE SECURITY

It is important to point out that the fourteen features of active and effective security programs emphasize that "one size does not fit all" and that they are not a cookie-cutter approach to effective implementation of security measures. There will be variability in security approaches and tactics among ESS facilities based on industry-specific circumstances and operating conditions. The fourteen features

- are sufficiently flexible to apply to all communication assets, regardless of size;
- incorporate the idea that active and effective security programs should have measurable goals and timelines; and
- allow flexibility for ESS industrial facilities to develop specific security approaches and tactics that are appropriate to industry-specific circumstances.

ESS facilities can differ in many ways, including:

- Number of supply sources
- Energy capacity
- Operation risk
- Location risk
- Security budget
- Spending priorities
- Political and public support
- Legal barriers
- Public vs. private ownership

ESS facilities should address security in an informed and systematic way, regardless of these differences. ESS facilities need to fully understand the specific, local circumstances and conditions under which they operate and develop a security program tailored to those conditions. The goal in identifying common features of active and effective security programs is to achieve consistency in security program outcomes among ESS facilities, while allowing for and encouraging facilities to develop utility-specific security approaches and tactics. The features are based on a comprehensive "security management layering system" approach that incorporates a combination of public involvement and awareness, partnerships, and physical, chemical, operational, and design controls to increase overall program performance. They address industry security in four functional categories: organization, operation, infrastructure, and external. These functional categories are discussed in greater detail below.

- Organizational: There is always something that can be done to improve security. Even when resources are limited, the simple act of increasing organizational attentiveness to security may reduce vulnerability and increase responsiveness. Preparedness itself can help deter attacks. The first step to achieving preparedness is to make security a part of the organizational culture so that it is in the day-to-day thinking of frontline employees, emergency responders, and management of every ESS facility in this country. To successfully incorporate security into "business as usual," there must be a strong commitment to security by organization leadership and by the supervising body, such as the board of stockholders. The following features address how a security culture can be incorporated into an organization.
- Operational: In addition to having a strong culture and awareness of security within an organization, an active and effective security program makes security part of operational activities, from daily operations, such as monitoring of physical access controls, to scheduled annual reassessments. ESS entities will often find that by implementing security into operations, they can also reap cost benefits and improve the quality or reliability of the energy service.
- Infrastructure: These recommendations advise utilities to address security in all elements of ESS infrastructure, from source to distribution and from processing to product delivery.
- External: Strong relationships with response partners and the public strengthen security and public confidence. Two of the recommended features of active and effective security programs address this need.

Fourteen Features

Feature 1. Make an explicit and visible commitment of the senior leadership to security.

ESS facilities should create an explicit, easily communicated, enterprise-wide commitment to security, which can be done through:

- Incorporating security into a utility-wide mission or vision statement, addressing the full scope of an active and effective security program—that is, protection of worker/public health, worker/public safety, and public confidence—and that is part of core day-to-day operations.
- Developing an enterprise-wide security policy or set of policies.

ESS entities should use the process of making a commitment to security as an opportunity to raise awareness of security throughout the organization, making the commitment visible to all employees and customers, and to help every facet of the enterprise to recognize the contribution they can make to enhancing security.

Feature 2. Promote security awareness throughout the organization.

The objective of a security culture should be to make security awareness a normal, accepted, and routine part of day-to-day operations. Examples of tangible efforts include:

- Conducting employee training
- Incorporating security into job descriptions
- Establishing performance standards and evaluations for security
- Creating and maintaining a security tip line and suggestion box for employees
- Making security a routine part of staff meetings and organization planning
- Creating a security policy

Feature 3. Assess vulnerabilities and periodically review and update vulnerability assessments to reflect changes in potential threats and vulnerabilities.

Because circumstances change, emergency services facilities should maintain their understanding and assessment of vulnerabilities as a "living document" and continually adjust their security enhancement and maintenance priorities. ESS facilities should consider their individual circumstances and establish and implement a schedule for review of their vulnerabilities.

Assessments should take place once every three to five years at a minimum. ESS facilities may be well served by doing assessments annually.

The basic elements of sound vulnerability assessments are:

- Characterization of the emergency services system, including its mission and objectives
- Identification and prioritization of adverse consequences to avoid

- Determination of critical assets that might be subject to malevolent acts that could result in undesired consequences
- Assessment of the likelihood (qualitative probability) of such malevolent acts from adversaries
- Evaluation of existing countermeasures
- Analysis of current risks and development of a prioritized plan for risk reduction

Feature 4. Identify security priorities and, on an annual basis, identify the resources dedicated to security programs and planned security improvements, if any.

Dedicated resources are important to ensure a sustained focus on security. Investment in security should be reasonable considering utilities' specific circumstances. In some circumstances, investment may be as simple as increasing the amount of time and attention that executives and managers give to security. Where threat potential or potential consequences are greater, greater investment likely is warranted.

This feature establishes the expectation that chemical industrial facilities should, through their annual capital, operations, maintenance, and staff resources plans, identify and set aside resources consistent with their specific identified security needs. Security priorities should be clearly documented and should be reviewed with utility executives at least once per year as part of the traditional budgeting process.

Feature 5. Identify managers and employees who are responsible for security and establish security expectations for all staff.

- Explicit identification of security responsibilities is important for the development of a security culture with accountability.
- At minimum, emergency services facilities should identify a single, designated individual responsible for overall security, even if other security roles and responsibilities will likely be dispersed throughout the organization.
- The number and depth of security-related roles will depend on a utility's specific circumstances.

Feature 6. Establish physical and procedural controls to restrict access to chemical industrial infrastructure to only those conducting authorized, official business, and to detect unauthorized physical intrusions.

Examples of physical access controls include fencing critical areas, locking gates and doors, and installing barriers at site access points. Monitoring for physical intrusion can include maintaining well-lighted facility perimeters, installing motion detectors, and utilizing intrusion alarms. The use of neighborhood watches, regular employee rounds, and arrangements with local police and fire departments can support identifying unusual activity in the vicinity of facilities.

Examples of procedural access controls include inventorying keys, changing access codes regularly, and requiring security passes to pass gates and access sensitive areas. In addition, utilities should establish the means to readily identify all employees, including contractors and temporary workers, with unescorted access to facilities.

Feature 7. Establish employee protocols for detection of contamination consistent with the recognized limitations in current contaminant detection, monitoring, and surveillance technology.

Until progress can be made in development of practical and affordable online contaminant monitoring and surveillance systems, most ESS facilities must use other approaches to contaminant monitoring and surveillance.

Many utilities already measure the above parameters (and many others) on a regular basis to control plant operations and confirm chemical mixture quality. More closely monitoring these parameters may create operational benefits for facilities that extend far beyond security, such as reducing operating costs and chemical usage. ESS facilities should also thoughtfully monitor customer complaints and improve connections with local public health networks to detect public health anomalies. Customer complaints and public health anomalies are an important way to detect potential contamination problems and other environmental quality concerns.

Feature 8. Define security-sensitive information; establish physical, electronic, and procedural controls to restrict access to security-sensitive information; detect unauthorized access; and ensure that information and communications systems will function during emergency response and recovery.

Protecting IT systems largely involves using physical hardening and procedural steps to limit the number of individuals with authorized access and to prevent access by unauthorized individuals. Examples of physical steps to harden SCADA and IT networks include installing and maintaining firewalls and screening the network for viruses. Examples of procedural steps include restricting remote access to data networks and safeguarding critical data through backups and storage in safe places. Utilities should strive for continuous operation of IT and telecommunications systems, even in the event of an attack, by providing uninterruptible power supply and backup systems, such as satellite phones.

In addition to protecting IT systems, security-sensitive information should be identified and restricted to the appropriate personnel. Security-sensitive information could be contained within:

- Facility maps and blueprints
- Operations details
- Hazardous material utilization and storage
- Tactical-level security program details
- Any other information on utility operations or technical details that could aid in planning or execution of an attack

Identification of security-sensitive information should consider all ways that utilities might use and make public information (e.g., many chemical industrial facilities may at times engage in competitive bidding processes for construction of new facilities or infrastructure). Finally, information critical to the continuity of day-to-day operations should be identified and backed up.

Feature 9. Incorporate security considerations into decisions about acquisition, repair, major maintenance, and replacement of physical infrastructure; include consideration of opportunities to reduce risk through physical hardening and adoption of inherently lower-risk design and technology options.

Prevention is a key aspect of enhancing security. Consequently, consideration of security issues should begin as early as possible in facility construction (i.e., it should be a factor in building plans and designs). However, to incorporate security considerations into design choices, chemical facilities need information about the types of security design approaches and equipment that are available and the performance of these designs and equipment in multiple dimensions. For example, ESS facilities would want to evaluate not just the way that a particular design might contribute to security but also how that design would affect the efficiency of day-to-day plant operations and worker safety.

Feature 10. Monitor available threat level information and escalate security procedures in response to relevant threats.

Monitoring threat information should be a regular part of a security program manager's job, and utility-, facility-, and region-specific threat levels and information should be shared with those responsible for security. As part of security planning, ESS facilities should develop systems to access threat information, establish procedures that will be followed in the event of increased industry or facility threat levels, and be prepared to put these procedures in place immediately so that adjustments are seamless. Involving local law enforcement and the FBI is critical.

ESS facilities should investigate what networks and information sources might be available to them locally and at the state and regional levels. If a utility cannot gain access to some information networks, attempts should be made to align with those who can and will provide effective information to the manufacturing facility.

Feature 11. Incorporate security considerations into emergency response and recovery plans, test and review plans regularly, and update plans to reflect changes in potential threats, physical infrastructure, chemical processing operations, critical interdependencies, and response protocols in partner organizations.

ESS facilities should maintain response and recovery plans as "living documents." In incorporating security considerations into their emergency response and recovery plans, chemical facilities should also be aware of the National Incident Management System (NIMS) guidelines, established by DHS, and of regional and local incident management commands and systems, which tend to flow from the national guidelines.

ESS facilities should consider their individual circumstances and establish, develop, and implement a schedule for review of emergency response and recovery plans. ESS facility plans should be thoroughly coordinated with emergency response and recovery planning in the larger community. As part of this coordination, a mutual-aid program should be established to arrange in advance for exchanging resources (personnel or physical assets) among agencies within a region in the event of an emergency or disaster that disrupts operation. Typically, the exchange of resources is based on a written formal and mutual agreement. For example, Florida's Water/Wastewater Agency Response Network (FlaWARN) deployed after Hurricane Katrina and allowed the new "utilities helping utilities" network to respond to urgent requests from Mississippi for help in bringing facilities back online after the hurricane.

The emergency response and recovery plans should be reviewed and updated as needed annually. This feature also establishes the expectation that chemical facilities should test or exercise their emergency response and recovery plans regularly.

Feature 12. Develop and implement strategies for regular, ongoing security-related communications with employees, response organizations, rate-setting organizations, and customers.

An active and effective security program should address protection of public health, public safety (including infrastructure), and public confidence. Emergency services facilities should create an awareness of security and an understanding of the rationale for their overall security management approach in the communities they reside in and/or serve.

Effective communication strategies consider key messages; who is best equipped/trusted to deliver the key messages; the need for message consistency, particularly during an emergency; and the best mechanisms for delivering messages and for receiving information and feedback from key partners. The key audiences for communication strategies are utility employees, response organizations, and customers.

Feature 13. Forge reliable and collaborative partnerships with the communities served, managers of critical interdependent infrastructure, response organizations, and other local utilities.

Effective partnerships build collaborative working relationships and clearly define roles and responsibilities so that people can work together seamlessly if an emergency should occur. It is important for ESS facilities within a region and in neighboring regions to collaborate and establish a mutual-aid program with neighboring utilities, response organizations, and sectors, such as the power sector, on which utilities rely or which they impact. Mutual-aid agreements provide for help from other organizations that is prearranged and can be accessed quickly and efficiently in the event of a terrorist attack or natural disaster. Developing reliable and collaborative partnerships involves reaching out to managers and key staff and other organizations to build reciprocal understanding and share information about the facility's security concerns and planning. Such efforts will maximize the efficiency and effectiveness of a mutual-aid program during an emergency response effort, as the organizations will be familiar with each other's circumstances and thus will be better able to serve each other.

It is also important for ESS facilities to develop partnerships with the communities and customers they serve. Partnerships help to build credibility within communities and establish public confidence in utility operations. People who live near emergency services facility structures can be the eyes and ears of the facility and can be encouraged to notice and report changes in operating procedures or other suspicious occurrences.

ESS facilities and public health organizations should establish formal agreements on coordination to ensure regular exchange of information between facilities and public health organizations and to outline roles and responsibilities during response to and recovery from an emergency. Coordination is important at all levels of the public health community—national public health, county health agencies, and health-care providers, such as hospitals.

Feature 14. Develop chemical facility–specific measures of security activities and achievements, and self-assess against these measures to understand and document program progress.

Although security approaches and tactics will be different depending on chemical utility–specific circumstances and operating conditions, we recommend that all emergency services facilities monitor and measure a number of common types of activities and achievements, including the existence of program policies and procedures, training, testing, and implementation schedules and plans.

The Fourteen Feature Matrix

Table 9.1 presents a matrix of recommended measures to assess the effectiveness of an ESS facility's security program. Each feature is grouped according to its functional category: organization, operation, infrastructure, or external.

Table 9.1. Fourteen Features of an Active and Effective Security Matrix

Features	*Checklist: Potential Measures of Progress*
Organizational Features	
Feature 1—Explicit commitment to security	Does a written, enterprise-wide security policy exist, and is the policy reviewed regularly and updated as needed?
Feature 2—Promotes security awareness	Are incidents reported in a timely way, and are lessons learned from incident responses reviewed and, as appropriate, incorporated into future utility security efforts?
Feature 5—Defined security roles and employee expectations	Are managers and employees who are responsible for security identified?
Operational Features	
Feature 3—Vulnerability assessment up to date	Are reassessments of vulnerabilities made after incidents, and are lessons learned and other relevant information incorporated into security practices?
Feature 4—Security resources and implementation priorities	Are security priorities clearly identified, and to what extent do security priorities have resources assigned to them?

Features	Checklist: Potential Measures of Progress
Feature 7—Contamination detection	Is there a protocol/procedure in place to identify and respond to suspected contamination events?
Feature 10—Threat-level-based protocols	Is there a protocol/procedure of responses that will be made if threat levels change?
Feature 11—Emergency response plan tested and up to date	Do exercises address the full range of threats—physical, cyber, and contamination—and is there a protocol/procedure to incorporate lessons learned from exercises and actual response into updates to emergency response and recovery plans?
Feature 14—Industry-specific measures and self-assessment	Does the utility perform self-assessment a least annually?
Infrastructure Features	
Feature 6—Intrusion detection and access control	To what extent are methods to control access to sensitive assets in place?
Feature 8—Information protection and continuity	Is there a procedure to identify and control security-sensitive information, is information correctly categorized, and how do control measures perform under testing?
Feature 9—Design and construction standards	Are security considerations incorporated into internal utility design and construction standards for new facilities/infrastructure and major maintenance projects?
External Features	
Feature 12—Communications	Is there a mechanism for utility employees, partners, and the community to notify the utility of suspicious occurrences and other security concerns?
Feature 13—Partnerships	Have reliable and collaborative partnerships with customers, managers of independent interrelated infrastructure, and response organizations been established?

Source: USEPA (2003).

Ultimately, the goal of implementing the fourteen security features (and all other security provisions) is to create a significant improvement in EES facilities on a national scale by reducing vulnerabilities and therefore the risk to public health from terrorist attacks and natural disasters. To create a sustainable effect, the communications sector as a whole must not only adopt and actively practice the features but also incorporate them into "business as usual."

REFERENCE

USEPA. (2003). *Active and Effective Water Security Programs.* http://cfpub.epa.gov/safewater/watersecurity/14 features.cfm, accessed June 2006.

Index